GENETICALLY ENGINEERED FOODS

Nancy Harris, *Book Editor*

Daniel Leone, *President*
Bonnie Szumski, *Publisher*
Scott Barbour, *Managing Editor*
Helen Cothran, *Senior Editor*

GREENHAVEN
PRESS ®

THOMSON
————————— ™
GALE

San Diego • Detroit • New York • San Francisco • Cleveland
New Haven, Conn. • Waterville, Maine • London • Munich

LIBRARY OF CONGRESS CATALOGING-IN-PUBLICATION DATA

Genetically engineered foods / Nancy Harris, book editor.
   p. cm. — (At issue)
Includes bibliographical references and index.
ISBN 0-7377-1787-4 (pbk. : alk. paper) — ISBN 0-7377-1786-6 (lib. : alk. paper)
   1. Genetically modified foods. I. Harris, Nancy, 1952– . II. At issue (San Diego, Calif.)
TP248.65.F66 G453 2004
641.3—dc21   2002035387

Contents

Introduction

The significance of genetic engineering is expressed in the words of Suzanne Wuerthele, an Environmental Protection Agency (EPA) toxicologist, who stated, "This is probably one of the most technologically powerful developments the world has ever seen. It's the biological equivalent of splitting the atom." Although humans have worked to improve plant breeding for thousands of years, genetic engineering, or genetic modification, is a new, even revolutionary, technology. Unlike conventional plant breeding, which involves the shifting of different forms of the same gene already present in a species' gene pool, genetic engineering usually involves the transfer of foreign genes, genes not previously present in a species' gene pool, into the organism. In genetic engineering, scientists transfer genetic information, or DNA, from one or more organisms across species boundaries into a host organism to create an entirely new genetically engineered organism. For example, the transfer of a gene from a bacterium into a corn plant can give the corn plant pesticide properties. In addition, other genes—genes responsible for promoting the correct functioning of the foreign gene and acting as markers or tags to identify the genetically engineered organism—are present alongside the gene for the desired characteristic. Genes for antibiotic resistance, for example, are commonly used as markers. Genetically engineered crops, therefore, are different from traditional plant varieties. While they do offer many benefits, they may also be more likely to have unpredictable physiological and biochemical effects.

Genetic engineering (GE) is the standard U.S. term for this new technology, also called recombinant DNA (rDNA) technology. In Europe, *genetically modified* (GM) is more commonly used to describe the technology, because this term translates more easily between different languages. The organism that is created through genetic engineering is called a *genetically modified organism* (GMO). The terms used to describe the foods produced from genetic engineering include *biotech foods*, *gene foods*, *bioengineered foods*, *gene-altered foods*, and *transgenic foods*.

In the 1970s scientists exchanged DNA between different strains of bacteria, producing the first genetically engineered organisms. Since that time, scientists have moved into the more complex field of genetically engineered plants and animals. Genes have been transferred between animal species, between plant species, and from animal species to plant species. Some genes can make an animal or plant grow faster or larger, or both. Canadian scientists found that flounder produce a type of antifreeze, and they transplanted a gene for this trait into salmon so that salmon can be farmed in colder climates. Many species of fish are genetically engineered to speed growth, to alter flesh quality, and to increase cold and disease resistance. In farm animals such as cattle, genes can be inserted to reduce the amount of fat in meat, to increase milk production,

5

and to increase superior cheese-making proteins in milk. In the area of plant crops, agricultural biotechnology can accelerate the process of transforming crops. Cotton, for example, was redesigned with a bacterial gene that allows the plant to produce its own pesticide, thus reducing the cost of spraying crops. Biotechnology has modified other plants to resist common diseases or to tolerate weed-killing herbicide sprays.

The development and advantages of GE foods

Foods produced using genetic modification arrived in U.S. markets in the mid-1990s, but a survey conducted in January 2001 by the Pew Initiative on Food and Biotechnology found that just 20 percent of U.S. consumers realized they had already eaten GE foods. According to the Grocery Manufacturers of America, an estimated 70 percent of the foods on grocery store shelves in the year 2000 were made or manufactured using GE crops. A large percentage of the corn and soybeans grown in the United States were from GE crops, which were processed into common food ingredients such as high fructose corn syrup. This syrup is used in many products, including soft drinks and the vegetable oil that is used to fry food such as french fries. In the spring of 2001, GE crops were grown on 82 million acres of U.S. farmland (up 18 percent from 2000), and 68 percent of all soybeans, 69 percent of all cotton, and 26 percent of all corn were genetically modified.

Multinational corporations such as Monsanto, Novartis AG, Aventis SA, and DuPont have spent hundreds of millions of dollars on biotechnology research in the past two decades, mostly to modify staples like corn, soybeans, and cotton, and make them easier and cheaper to plant and grow—and therefore more profitable for agribusiness companies. As Americans became aware of GE foods, antibiotech sentiment emerged in the United States. As a result, in the spring of 2000, the top U.S. biotech companies, normally staunch competitors, formed the Council for Biotechnology Information to launch a $50-million-a-year public relations campaign. Proponents of genetic engineering claim that GE foods provide advantages to human health. In addition to enhanced nutritional value and increased yields from GE seeds, genetic modification can produce cooking oils with less saturated fat, soybean oil with high levels of vitamin E, and strawberries and tomatoes with a cancer-fighting chemical. Medical benefits from genetic engineering include edible vaccines such as bananas spliced with genes from a hepatitis B virus and "golden rice," which contains a daffodil gene that provides vitamin K, essential in promoting healthier eyes in children. (Thousands of children suffer blindness every year because of a lack of vitamin K.) In Japan, genetic engineering has experimentally removed a protein from rice that provokes allergic reactions, and similar work is being done on peanuts, to which some people may have violent allergic reactions.

The safety of GE foods to human health

"The use of rDNA biotechnology in itself has no impact on the safety of foods derived using these techniques," states an April 2001 report by the Institute of Food Technologists, a scientific society concerned with food science and technology. The report further states, "The appropriate federal agencies have reviewed all the rDNA biotechnologically-derived foods on the market to ensure their safety." Nevertheless, many American consumers became aware of the market presence of GE foods in September 2000, when StarLink corn, a bioengineered corn that was not cleared for human consumption, was discovered in taco shells and other corn products. As a result, concerns arose in the United States about the safety of GE foods, and those concerns have persisted.

Some scientists believe that genetic engineering dangerously tampers with the most fundamental natural components of life and that genetic engineering is scientifically unsound. In addition, they believe that when scientists transfer genes into a new organism, the results could be unexpected and dangerous. No long-term studies have been done to determine what effects many of the commonly consumed GE foods might have on human health. Critics accuse GE supporters of assuming that, because the modified foods have been on the market for as long as they have with no apparent ill effects, they must be safe. However, some research and case studies suggest that GE foods may cause adverse health effects, including increased exposure to allergens, elevated cancer risks, and increased resistance to antibiotics as a result of the common practice of using antibiotic-resistant genes as marker genes in these foods. In 1999, a number of independent scientists, along with the British Medical Association, called for a moratorium on GE foods until further research is completed. Food and Drug Administration (FDA) scientists have warned that genetic engineering could result in undesirable changes in the level of nutrients because fresh-looking foods may actually be weeks old and therefore be nutritionally deficient. FDA scientist Edwin Mathews warned that the levels of toxins in GE foods could be higher than in normal foods and that GE foods might produce new toxic substances.

Controversy over regulations

The United States has no mandatory labeling of GE foods to inform consumers whether or not the foods they are buying have been genetically engineered. In addition, the United States has no regulations requiring human testing of GE crops. The only requirement is that producers report to the FDA the food's nutritional properties and the source of the genes used to genetically alter the food. Biotech companies contend that the FDA should ensure food safety, but the FDA claims that ultimately it is the responsibility of biotech companies to ensure that their foods are safe.

In 1990, the Bush administration issued a set of regulatory principles aimed at smoothing the way for the speedy development of biotechnology products in the United States (and to help large biotech corporations in the world market). The first principle stated, "Federal government regulatory oversight should focus on the characteristics and risks of the biotechnology product—not the process by which it was created." This meant that FDA, EPA, and U.S. Department of Agriculture (USDA) scientists were not to single out GE foods for review simply because these foods had been developed using a new technology. The reason for this decision was the apparent absence of any information showing differences between the effects of genetic engineering and those of traditional animal and plant breeding. However, internal documents made public in 1999 revealed that many FDA scientists disagreed with this decision, including Louis Pribyl, who said, "There is a profound difference between the types of unexpected effects from traditional breeding and genetic engineering." One major difference is that genetic engineering crosses the natural biological boundaries between animal and plant species, whereas conventional animal and plant breeding does not.

Although some researchers believe that genetically engineered foods will bring about beneficial changes for humankind, others question whether scientists, regulators, and consumers have enough information about GE foods to ensure their safety. *At Issue: Genetically Engineered Foods* provides different perspectives on this highly controversial technology.

1

Genetically Engineered Foods: An Overview

Cindy Maynard

Cindy Maynard is a registered dietician and a health and medical writer and editor for Current Health 2.

Genetically modified foods have been creating a "quiet revolution" in the American market for the past several years. By transferring genes from one organism into another across species boundaries, scientists are able to produce new foods—foods that have raised a host of controversial questions. Along with the apparent benefits of increased disease resistance in food crops, higher crop yields, and possible medical benefits come questions of safety to human health and to the environment. These safety concerns also raise the issue of whether genetically modified foods should be labeled as such. Conflicting scientific reports suggest that these questions may not be easily answered.

Genetically modified foods are increasingly a part of our diet. Not everyone feels this is a good thing.

If you've gobbled down a crispy taco, a bowl of cornflakes, a baked potato, or cheese pizza lately, chances are you've eaten genetically engineered foods. The British and anti-biotech groups are dubbing these foods "Frankenfood."

No, it's not sci-fi fantasy.

There has been a quiet revolution occurring in the American diet in the past few years. But lately, it hasn't been so quiet. It's now estimated that approximately 70 percent of the foods we eat have some genetically engineered component, and the figure seems to be growing. Is it a problem? Are you in any danger? It depends on whom you ask.

How genetic engineering works

For thousands of years, farmers have bred animals and crops to produce better outcomes in terms of yield, pest or drought resistance, or to speed

Cindy Maynard, "Biotech at the Table," *Current Health 2*, vol. 27, November 2000, p. 22. Copyright © 2000 by Weekly Reader Corporation. Reproduced by permission.

up ripening. How is using biotechnology different? Biotech applies biological methods to improve plant or animal characteristics.

This brave new world of genetically modified organisms (GMOs) is the result of scientists being able to pick a desired gene from one organism—a bacterium, plant, or animal—and transplant it into the DNA of another organism to achieve the desired results. Some examples: a stalk of corn that breeds a new generation of pest-resisting corn, a potato that becomes resistant to the potato beetle, or a strawberry able to fend off frost. Another example is the soybean. A soy DNA is spliced with petunia DNA to produce a plant engineered to survive otherwise toxic doses of herbicide.

"Traditional crossbreeding requires the mixing of thousands of genes between two plants in the hopes of getting the desired trait. With modern biotechnology, you can choose the characteristic you want and add that single feature to a seed," says Felicia Busch, an American Dietetic Association (ADA) spokesperson and author of *From Antioxidants to Zucchini*.

This ability to manipulate genes is having a profound impact on American agriculture. Consider these basic possibilities:

• Using biotechnology can make a crop more resistant to pests, herbicides, or disease.

• A plant can be made to produce more of a nutrient, such as protein or a particular vitamin.

• The crop can potentially increase its yield.

The first biotech version of a major crop hit the market back in 1994 with a new breed of tomato. By 1998, 25 percent of corn and 38 percent of soybeans grown in the United States were genetically modified. Now it's virtually impossible not to eat these foods because they have found their way into the agricultural mainstream as ingredients in processed foods.

There has been a quiet revolution occurring in the American diet in the past few years.

About 60 percent of processed foods are made with soy and corn derivatives, some of which end up in our salads, corn chips, oils, french fries, snacks, or side dishes. Biologically engineered corn is being used in many other familiar foods, such as breakfast cereals and taco shells. It is also used to make corn syrup, an ingredient in soft drinks, baked goods, and candies. Genetically altered soybeans are used in many cooking oils, soy sauce, candies, and margarine. Other products might include the tomatoes that go into your favorite spaghetti sauce or pizza topping.

Controversy over bioengineered foods

There is a great deal of controversy regarding bioengineered foods. The federal government and industry generally defend GMOs. When farmers can increase disease resistance in their crops, the yield improves. These groups claim that by producing more of these crops, millions of people in the Third World can be saved from starvation or blindness caused by malnutrition. Some food industries are betting heavily on the future with genetically modified foods, hoping that crops will be manipulated to create more nutritious foods or even foods that could potentially deliver medication just by their consumption.

But some critics claim such tinkering can go haywire as the industry shuffles genes around. Opponents such as the environmental organization Greenpeace and some scientists are concerned about the dangers and long-term health problems of dabbling in gene splicing. Politically, the subject has caused problems for the United States in that critics in Europe and other countries are refusing to buy genetically modified foods.

Scientific reports vary, and clear answers may not come quickly.

There are health and environmental concerns such as:

• Genetically altered foods may have a long-term impact on other crops or species and the environment. Will there be unpredictable combinations of genetic material or damage to the ecosystem?

• There is the possibility that GMOs produced with antibiotic-resistant genes might create resistance to antibiotics in humans who eat products containing them.

• Genetically modified foods may trigger the immune system to produce an allergic reaction; however, new GMO foods are tested for potential allergens. A recent example is when a company put a Brazil nut gene into a soybean: People who are allergic to nuts could potentially develop an allergic reaction by eating the genetically altered soybean. Because of this problem, this product was never marketed.

• Should genetically altered foods be labeled so consumers can decide whether or not they want to eat them? The U.S. Food and Drug Administration (FDA) is currently reviewing this.

Organizations responsible for GMO food regulation

Organizations involved in regulating genetically modified foods include:

FDA: They focus on the safety of a new plant, animal, or product rather than the mechanism by which it was developed.

EPA: The U.S. Environmental Protection Agency establishes regulations and registration requirements based upon the toxicity and environmental impact of new pesticide products or plants with pesticidal traits.

USDA: The U.S. Department of Agriculture develops policy and regulations, and reviews all environmental releases of genetically engineered organisms, plants, and animals.

"Researchers must now explore the long-term effects of using genetically modified organisms, but the prospects are still promising. After all, without GMOs we wouldn't have lifesaving human insulin for Type I diabetics," states ADA [American Diabetes Association] official Christina Ferroli of the Consumer and Family Sciences Department of Purdue Extension—Marion County, Indianapolis.

But Michael Jacobson, executive director of the Center for Science in the Public Interest, offers a different slant. "Genetic engineering is a powerful tool if used properly," says Jacobson. "But the government needs to tighten its regulation of genetically modified foods, and the FDA needs to review the safety of every such food before it hits the market. Genetically modified foods should be clearly labeled as such."

Scientific reports vary, and clear answers may not come quickly. For now, here are some recommendations that may help you evaluate the pros and cons.

Concerning the safety of GE foods

In April 2000, the National Academy of Sciences generally endorsed the safety of bioengineered plants, but advised the U.S. government to do the following:

• Make sure plants that have been bioengineered to produce their own pesticides don't harm insects that are not pests. For example, re-searchers have reported that monarch butterfly caterpillars exposed to pollen from genetically engineered corn died or developed abnormally. Further research is needed.

• Develop better tests to identify allergy-causing proteins in geneti-cally modified foods.

• Prevent genes that have been introduced into crops from spreading to wild relatives of those crops, which could create herbicide-resistant weeds or "super weeds."

• Have mandatory labeling of genetically modified products.

• Conduct animal testing to study the potential toxicity of the prod-ucts and their impact on human growth and development.

Those who are strongly skeptical of biologically engineered foods can purchase organic foods. Currently, this is the only way to avoid geneti-cally modified foods.

As Jacobson sums it up: "It's likely that the public will hotly debate biotechnology for years to come."

2

Exploring the Safety and Ethics of Genetically Engineered Foods

Michael Pollan

Michael Pollan was editor-at-large for Harper's Magazine, *contributing editor for the* New York Times Magazine, *and is the author of* The Botany of Desire: A Plant's Eye View of the World *and* Second Nature: A Gardener's Education.

Many uncertainties still exist in the area of genetically engineered food safety for humans and for the environment. Will insects develop resistance to crops with built-in insecticides, creating a "biological pollution"? Is it safe for these foods to continue to be put on the market when scientists say there are still many unknowns in the field of genetic engineering? And why have Europeans so strongly resisted genetically engineered (GE) foods? There are also questions about the regulation of these foods. Are the Food and Drug Administration and Environmental Protection Agency, which are the food and pesticide regulating agencies, doing an adequate job, and why have the controversial GE foods not been labeled to inform consumers about them? Conventional farmers may enthusiastically accept the new biotechnology for reducing the widespread use of herbicides and pesticides while questioning whether it is worth the price of having big business tighten its "noose" around their necks. Organic growers, with their more nature-friendly farming practices, consider genetic engineering synthetic, questionable, and not what their customers want. In the end, all the unknowns foster an environment of uncertainty concerning the safety of genetically engineered foods.

Today I planted something new in my vegetable garden—something very new, as a matter of fact. It's a potato called the New Leaf Superior, which has been genetically engineered—by Monsanto, the chemical giant recently turned "life sciences" giant—to produce its own insecticide. This it can do in every cell of every leaf, stem, flower, root and (here's the creepy part) spud. The scourge of potatoes has always been the Colorado potato beetle, a handsome and voracious insect that can pick a plant clean of its leaves virtually overnight. Any Colorado potato beetle that takes so much as a nibble of my New Leafs will supposedly keel over and die, its digestive tract pulped, in effect, by the bacterial toxin manufactured in the leaves of these otherwise ordinary Superiors. (Superiors are the thin-skinned white spuds sold fresh in the supermarket.) You're probably wondering if I plan to eat these potatoes, or serve them to my family. That's still up in the air; it's only the first week of May, and harvest is a few months off.

Certainly my New Leafs are aptly named. They're part of a new class of crop plants that is rapidly changing the American food chain. This year [1998], the fourth year that genetically altered seed has been on the market, some 45 million acres of American farmland have been planted with biotech crops, most of it corn, soybeans, cotton and potatoes that have been engineered to either produce their own pesticides or withstand herbicides. Though Americans have already begun to eat genetically engineered potatoes, corn and soybeans, industry research confirms what my own informal surveys suggest: hardly any of us knows it. The reason is not hard to find. The biotech industry, with the concurrence of the Food and Drug Administration, has decided we don't need to know it, so biotech foods carry no identifying labels. In a dazzling feat of positioning, the industry has succeeded in depicting these plants simultaneously as the linchpins of a biological revolution—part of a "new agricultural paradigm" that will make farming more sustainable, feed the world and improve health and nutrition—and, oddly enough, as the same old stuff, at least so far as those of us at the eating end of the food chain should be concerned.

The introduction into a plant of genes transported not only across species but whole phyla means that the wall of that plant's essential identity—its irreducible wildness, you might say—has been breached.

This convenient version of reality has been roundly rejected by both consumers and farmers across the Atlantic. [In the summer of 1998], biotech food emerged as the most explosive environmental issue in Europe. Protesters have destroyed dozens of field trials of the very same "frankenplants" (as they are sometimes called) that we Americans are already serving for dinner, and throughout Europe the public has demanded that biotech food be labeled in the market.

By growing my own transgenic crop—and talking with scientists and farmers involved with biotech—I hoped to discover which of us was crazy. Are the Europeans overreacting, or is it possible that we've been underreacting to genetically engineered food?

After digging two shallow trenches in my garden and lining them with compost, I untied the purple mesh bag of seed potatoes that Monsanto had sent and opened up the Grower Guide tied around its neck.

(Potatoes, you may recall from kindergarten experiments, are grown not from seed but from the eyes of other potatoes.) The guide put me in mind not so much of planting potatoes as booting up a new software release. By "opening and using this product," the card stated, I was now "licensed" to grow these potatoes, but only for a single generation; the crop I would water and tend and harvest was mine, yet also not mine. That is, the potatoes I will harvest come August are mine to eat or sell, but their genes remain the intellectual property of Monsanto, protected under numerous United States patents, including Nos. 5,196,525, 5,164,316, 5,322,938 and 5,352,605. Were I to save even one of them to plant next year—something I've routinely done with potatoes in the past—I would be breaking Federal law. The small print in the Grower Guide also brought the news that my potato plants were themselves a pesticide, registered with the Environmental Protection Agency.

There's no way of telling where in the genome the new DNA will land.

If proof were needed that the intricate industrial food chain that begins with seeds and ends on our dinner plates is in the throes of profound change, the small print that accompanied my New Leaf will do. That food chain has been unrivaled for its productivity—on average, a single American farmer today grows enough food each year to feed 100 people. But this accomplishment has come at a price. The modern industrial farmer cannot achieve such yields without enormous amounts of chemical fertilizer, pesticide, machinery and fuel, a set of capital-intensive inputs, as they're called, that saddle the farmer with debt, threaten his health, erode his soil and destroy its fertility, pollute the ground water and compromise the safety of the food we eat.

We've heard all this before, of course, but usually from environmentalists and organic farmers; what is new is to hear the same critique from conventional farmers, government officials and even many agribusiness corporations, all of whom now acknowledge that our food chain stands in need of reform. Sounding more like Wendell Berry [a conservationist, novelist, and poet] than the agribusiness giant it is, Monsanto declared in its most recent annual report that "current agricultural technology is not sustainable."

What is supposed to rescue the American food chain is biotechnology—the replacement of expensive and toxic chemical inputs with expensive but apparently benign genetic information: crops that, like my New Leafs, can protect themselves from insects and disease without being sprayed with pesticides. With the advent of biotechnology, agriculture is entering the information age, and more than any other company, Monsanto is positioning itself to become its Microsoft, supplying the proprietary "operating systems"—the metaphor is theirs—to run this new generation of plants.

There is, of course, a second food chain in America: organic agriculture. And while it is still only a fraction of the size of the conventional food chain, it has been growing in leaps and bounds—in large part because of concerns over the safety of conventional agriculture. Organic farmers have been among biotechnology's fiercest critics, regarding crops like my New Leafs as inimical to their principles and, potentially, a threat to their survival. That's because Bt, the bacterial toxin produced in my New Leafs (and in many other biotech plants) happens to be the same insecticide organic growers have relied on for decades. Instead of being flattered by the imitation, however, organic farmers are up in arms: the widespread use of Bt in biotech crops is likely to lead to insect resistance, thus robbing organic growers of one of their most critical tools; that is, Monsanto's version of sustainable agriculture may threaten precisely those farmers who pioneered sustainable farming.

Pushing up shoots

After several days of drenching rain, the sun appeared on May 15, and so did my New Leafs. A dozen deep-green shoots pushed up out of the soil and commenced to grow—faster and more robustly than any of the other potatoes in my garden. Apart from their vigor, though, my New Leafs looked perfectly normal. And yet as I watched them multiply their lustrous dark-green leaves those first few days, eagerly awaiting the arrival of the first doomed beetle, I couldn't help thinking of them as existentially different from the rest of my plants.

All domesticated plants are in some sense artificial—living archives of both cultural and natural information that we in some sense "design." A given type of potato reflects the values we've bred into it—one that has been selected to yield long, handsome french fries or unblemished round potato chips is the expression of a national food chain that likes its potatoes highly processed. At the same time, some of the more delicate European fingerlings I'm growing alongside my New Leafs imply an economy of small market growers and a taste for eating potatoes fresh. Yet all these qualities already existed in the potato, somewhere within the range of genetic possibilities presented by Solanum tuberosum. Since distant species in nature cannot be crossed, the breeder's art has always run up against a natural limit of what a potato is willing, or able, to do. Nature, in effect, has exercised a kind of veto on what culture can do with a potato.

My New Leafs are different. Although Monsanto likes to depict biotechnology as just another in an ancient line of human modifications of nature going back to fermentation, in fact genetic engineering overthrows the old rules governing the relationship of nature and culture in a plant. For the first time, breeders can bring qualities from anywhere in nature into the genome of a plant—from flounders (frost tolerance), from viruses (disease resistance) and, in the case of my potatoes, from Bacillus thuringiensis, the soil bacterium that produces the organic insecticide known as Bt. The introduction into a plant of genes transported not only across species but whole phyla means that the wall of that plant's essential identity—its irreducible wildness, you might say—has been breached.

But what is perhaps most astonishing about the New Leafs coming up in my garden is the human intelligence that the inclusion of the Bt gene represents. In the past, that intelligence resided outside the plant, in the mind of the organic farmers who deployed Bt (in the form of a spray) to manipulate the ecological relationship of certain insects and a certain bacterium as a way to foil those insects. The irony about the New Leafs is that the cultural information they encode happens to be knowledge that resides in the heads of the very sort of people—that is, organic growers—who most distrust high technology.

One way to look at biotechnology is that it allows a larger portion of human intelligence to be incorporated into the plant itself. In this sense, my New Leafs are just plain smarter than the rest of my potatoes. The others will depend on my knowledge and experience when the Colorado potato beetles strike; the New Leafs, knowing what I know about bugs and Bt, will take care of themselves. So while my biotech plants might seem like alien beings, that's not quite right. They're more like us than like other plants because there's more of us in them.

At the lab

To find out how my potatoes got that way, I traveled to suburban St. Louis in early June. My New Leafs are clones of clones of plants that were first engineered seven years ago in Monsanto's $150 million research facility, a long, low-slung brick building on the banks of the Missouri that would look like any other corporate complex were it not for the 26 greenhouses that crown its roof like shimmering crenellations of glass.

Dave Stark, a molecular biologist and co-director of Naturemark, Monsanto's potato subsidiary, escorted me through the clean rooms where potatoes are genetically engineered. Technicians sat at lab benches before petri dishes in which fingernail-size sections of potato stem had been placed in a nutrient mixture. To this the technicians added a solution of agrobacterium, a disease bacterium whose modus operandi is to break into a plant cell's nucleus and insert some of its own DNA. Essentially, scientists smuggle the Bt gene into the agrobacterium's payload, and then the bacterium splices it into the potato's DNA. The technicians also add a "marker" gene, a kind of universal product code that allows Monsanto to identify its plants after they leave the lab.

"We have such a miserably poor understanding of how the organism develops from its DNA that I would be surprised if we don't get one rude shock after another."

A few days later, once the slips of potato stem have put down roots, they're moved to the potato greenhouse up on the roof. Here, Glenda DeBrecht, a horticulturist, invited me to don latex gloves and help her transplant pinky-size plantlets from their petri dish to small pots. The whole operation is performed thousands of times, largely because there is so much uncertainty about the outcome. There's no way of telling where in the genome the new DNA will land, and if it winds up in the wrong place, the new gene won't be expressed (or it will be poorly expressed) or the plant may be a freak. I was struck by how the technology could at once be astoundingly sophisticated and yet also a shot in the genetic dark.

"There's still a lot we don't understand about gene expression," Stark acknowledged. A great many factors influence whether, or to what extent, a new gene will do what it's supposed to, including the environment. In one early German experiment, scientists succeeded in splicing the gene for redness into petunias. All went as planned until the weather turned hot and an entire field of red petunias suddenly and inexplicably lost their pigment. The process didn't seem nearly as simple as Monsanto's cherished software metaphor would suggest.

When I got home from St. Louis, I phoned Richard Lewontin, the Harvard geneticist, to ask him what he thought of the software metaphor. "From an intellectual-property standpoint, it's exactly right," he said. "But it's a bad one in terms of biology. It implies you feed a program into a machine and get predictable results. But the genome is very noisy. If my computer made as many mistakes as an organism does"—in interpreting its DNA, he meant—"I'd throw it out."

I asked him for a better metaphor. "An ecosystem," he offered, "you can always intervene and change something in it, but there's no way of knowing what all the downstream effects will be or how it might affect the environment. We have such a miserably poor understanding of how the organism develops from its DNA that I would be surprised if we don't get one rude shock after another."

Uncertainty of biotech agriculture

My own crop was thriving when I got home from St. Louis; the New Leafs were as big as bushes, crowned with slender flower stalks. Potato flowers are actually quite pretty, at least by vegetable standards—five-petaled pink stars with yellow centers that give off a faint rose perfume. One sultry afternoon I watched the bumblebees making their lazy rounds of my potato blossoms, thoughtlessly powdering their thighs with yellow pollen grains before lumbering off to appointments with other blossoms, others species.

Uncertainty is the theme that unifies much of the criticism leveled against biotech agriculture by scientists and environmentalists. By planting millions of acres of genetically altered plants, we have introduced something novel into the environment and the food chain, the consequences of which are not—and at this point, cannot be—completely understood. One of the uncertainties has to do with those grains of pollen bumblebees are carting off from my potatoes. That pollen contains Bt genes that may wind up in some other, related plant, possibly conferring a new evolutionary advantage on that species. "Gene flow," the scientific term for this phenomenon, occurs only between closely related species, and since the potato evolved in South America, the chances are slim that my Bt potato genes will escape into the wilds of Connecticut. (It's interesting to note that while biotechnology depends for its power on the ability to move genes freely among species and even phyla, its environmental safety depends on the very opposite phenomenon: on the integrity of species in nature and their rejection of foreign genetic material.)

Yet what happens if and when Peruvian farmers plant Bt potatoes? Or when I plant a biotech crop that does have local relatives? A study reported in *Nature* [in September 1998] found that plant traits introduced by genetic engineering were more likely to escape into the wild than the same traits introduced conventionally.

Biological pollution

Andrew Kimbrell, director of the Center for Technology Assessment in Washington, told me he believes such escapes are inevitable. "Biological pollution will be the environmental nightmare of the 21st century," he said when I reached him by phone. "This is not like chemical pollution—an oil spill—that eventually disperses. Biological pollution is an entirely different model, more like a disease. Is Monsanto going to be held legally responsible when one of its transgenes creates a superweed or resistant insect?"

Kimbrell maintains that because our pollution laws were written before the advent of biotechnology, the new industry is being regulated under an ill-fitting regime designed for the chemical age. Congress has so far passed no environmental law dealing specifically with biotech. Monsanto, for its part, claims that it has thoroughly examined all the potential environmental and health risks of its biotech plants, and points out that three regulatory agencies—the U.S.D.A., the E.P.A. and the F.D.A.—have signed off on its products. Speaking of the New Leaf, Dave Stark told me, "This is the most intensively studied potato in history."

Insect resistance

Significant uncertainties remain, however. Take the case of insect resistance to Bt, a potential form of "biological pollution" that could end the effectiveness of one of the safest insecticides we have—and cripple the organic farmers who depend on it. The theory, which is now accepted by most entomologists, is that Bt crops will add so much of the toxin to the environment that insects will develop resistance to it. Until now, resistance hasn't been a worry because the Bt sprays break down quickly in sunlight and organic farmers use them only sparingly. Resistance is essentially a form of co-evolution that seems to occur only when a given pest population is threatened with extinction; under that pressure, natural selection favors whatever chance mutations will allow the species to change and survive.

Working with the E.P.A., Monsanto has developed a "resistance-management plan" to postpone that eventuality. Under the plan, farmers who plant Bt crops must leave a certain portion of their land in non-Bt crops to create "refuges" for the targeted insects. The goal is to prevent the first Bt-resistant Colorado potato beetle from mating with a second resistant bug, unleashing a new race of superbeetles. The theory is that when a Bt-resistant bug does show up, it can be induced to mate with a susceptible bug from the refuge, thus diluting the new gene for resistance.

But a lot has to go right for Mr. Wrong to meet Miss Right. No one is sure how big the refuges need to be, where they should be situated or whether the farmers will cooperate (creating havens for a detested pest is counter-intuitive, after all), not to mention the bugs. In the case of potatoes, the E.P.A. has made the plan voluntary and lets the companies themselves implement it; there are no E.P.A. enforcement mechanisms. Which is why most of the organic farmers I spoke to dismissed the regulatory scheme as window dressing.

Monsanto executives offer two basic responses to criticism of their Bt crops. The first is that their voluntary resistance-management plans will

work, though the company's definition of success will come as small con-
solation to an organic farmer: Monsanto scientists told me that if all goes
well, resistance can be postponed for 30 years. (Some scientists believe it
will come in three to five years.) The second response is more troubling.
In St. Louis, I met with Jerry Hjelle, Monsanto's vice president for regulatory affairs. Hjelle told me that resistance should not unduly concern us
since "there are a thousand other Bt's out there"—other insecticidal proteins. "We can handle this problem with new products," he said. "The
critics don't know what we have in the pipeline."

And then Hjelle uttered two words that I thought had been expunged
from the corporate vocabulary a long time ago: "Trust us."

---

*"Biological pollution will be the environmental
nightmare of the 21st century."*

Trust is a key to the success of biotechnology in the marketplace, and
while I was in St. Louis, I asked Hjelle and several of his colleagues why
they thought the Europeans were resisting biotech food. Austria, Luxem-
bourg and Norway, risking trade war with the United States, have refused
to accept imports of genetically altered crops. Activists in England have
been staging sit-ins and "decontaminations" in biotech test fields. A
group of French farmers broke into a warehouse and ruined a shipment
of biotech corn seed by urinating on it. The Prince of Wales, who is an ar-
dent organic gardener, waded into the biotech debate [in June 1998],
vowing in a column in *The Daily Telegraph* that he would never eat, or
serve to his guests, the fruits of a technology that "takes mankind into
realms that belong to God and to God alone."

Monsanto executives are quick to point out that mad cow disease has
made Europeans extremely sensitive about the safety of their food chain
and has undermined confidence in their regulators. "They don't have a
trusted agency like the F.D.A. looking after the safety of their food supply," said Phil Angell, Monsanto's director of corporate communications.
Over the summer [of 1998], Angell was dispatched repeatedly to Europe
to put out the P.R. fires; some at Monsanto worry these could spread to
the United States.

Checking with the authorities

I checked with the F.D.A. to find out exactly what had been done to in-
sure the safety of this potato. I was mystified by the fact that the Bt toxin
was not being treated as a "food additive" subject to labeling, even
though the new protein is expressed in the potato itself. The label on a
bag of biotech potatoes in the supermarket will tell a consumer all about
the nutrients they contain, even the trace amounts of copper. Yet it is
silent not only about the fact that those potatoes are the product of genetic engineering but also about their containing an insecticide.

At the F.D.A., I was referred to James Maryanski, who oversees
biotech food at the agency. I began by asking him why the F.D.A. didn't
consider Bt a food additive. Under F.D.A. law, any novel substance added

to a food must—unless it is "generally regarded as safe" ("GRAS," in F.D.A. parlance)—be thoroughly tested and if it changes the product in any way, must be labeled.

"That's easy," Maryanski said. "Bt is a pesticide, so it's exempt" from F.D.A. regulation. That is, even though a Bt potato is plainly a food, for the purposes of Federal regulation it is not a food but a pesticide and therefore falls under the jurisdiction of the E.P.A.

Yet even in the case of those biotech crops over which the F.D.A. does have jurisdiction, I learned that F.D.A. regulation of biotech food has been largely voluntary since 1992, when Vice President Dan Quayle issued regulatory guidelines for the industry as part of the Bush Administration's campaign for "regulatory relief." Under the guidelines, new proteins engineered into foods are regarded as additives (unless they're pesticides), but as Maryanski explained, "the determination whether a new protein is GRAS can be made by the company." Companies with a new biotech food decide for themselves whether they need to consult with the F.D.A. by following a series of "decision trees" that pose yes or no questions like this one: "Does . . . the introduced protein raise any safety concern?"

Since my Bt potatoes were being regulated as a pesticide by the E.P.A. rather than as a food by the F.D.A., I wondered if the safety standards are the same. "Not exactly," Maryanski explained. The F.D.A. requires "a reasonable certainty of no harm" in a food additive, a standard most pesticides could not meet. After all, "pesticides are toxic to something," Maryanski pointed out, so the E.P.A. instead establishes human "tolerances" for each chemical and then subjects it to a risk-benefit analysis.

When I called the E.P.A. and asked if the agency had tested my Bt potatoes for safety as a human food, the answer was . . . not exactly.

When I called the E.P.A. and asked if the agency had tested my Bt potatoes for safety as a human food, the answer was . . . not exactly. It seems the E.P.A. works from the assumption that if the original potato is safe and the Bt protein added to it is safe, then the whole New Leaf package is presumed to be safe. Some geneticists believe this reasoning is flawed, contending that the process of genetic engineering itself may cause subtle, as yet unrecognized changes in a food.

The original Superior potato is safe, obviously enough, so that left the Bt toxin, which was fed to mice, and they "did fine, had no side effects," I was told. I always feel better knowing that my food has been poison-tested by mice, though in this case there was a small catch: the mice weren't actually eating the potatoes, not even an extract from the potatoes, but rather straight Bt produced in a bacterial culture.

So are my New Leafs safe to eat? Probably, assuming that a New Leaf is nothing more than the sum of a safe potato and a safe pesticide, and further assuming that the E.P.A.'s idea of a safe pesticide is tantamount to a safe food. Yet I still had a question. Let us assume that my potatoes are a pesticide—a very safe pesticide. Every pesticide in my garden shed—including the Bt sprays—carries a lengthy warning label. The label on my bottle of Bt says, among other things, that I should avoid inhaling the spray or getting it in an open wound. So if my New Leaf potatoes contain an E.P.A.-registered pesticide, why don't they carry some such label?

Maryanski had the answer. At least for the purposes of labeling, my New Leafs have morphed yet again, back into a food: the Food, Drug and Cosmetic Act gives the F.D.A. sole jurisdiction over the labeling of plant foods, and the F.D.A. has ruled that biotech foods need be labeled only if they contain known allergens or have otherwise been "materially" changed.

But isn't turning a potato into a pesticide a material change?

It doesn't matter. The Food, Drug and Cosmetic Act specifically bars the F.D.A. from including any information about pesticides on its food labels.

I thought about Maryanski's candid and wondrous explanations the next time I met Phil Angell, who again cited the critical role of the F.D.A. in assuring Americans that biotech food is safe. But this time he went even further. "Monsanto should not have to vouchsafe the safety of biotech food," he said. "Our interest is in selling as much of it as possible. Assuring its safety is the F.D.A.'s job."

Meeting the beetles

My Colorado potato beetle vigil came to an end the first week of July, shortly before I went to Idaho to visit potato growers. I spied a single mature beetle sitting on a New Leaf leaf; when I reached to pick it up, the beetle fell drunkenly to the ground. It had been sickened by the plant and would soon be dead. My New Leafs were working.

From where a typical American potato grower stands, the New Leaf looks very much like a godsend. That's because where the typical potato grower stands is in the middle of a bright green field that has been doused with so much pesticide that the leaves of his plants wear a dull white chemical bloom that troubles him as much as it does the rest of us. Out there, at least, the calculation is not complex: a product that promises to eliminate the need for even a single spraying of pesticide is, very simply, an economic and environmental boon.

Meeting with the farmers

No one can make a better case for a biotech crop than a potato farmer, which is why Monsanto was eager to introduce me to several large growers. Like many farmers today, the ones I met feel trapped by the chemical inputs required to extract the high yields they must achieve in order to pay for the chemical inputs they need. The economics are daunting: a potato farmer in south-central Idaho will spend roughly $1,965 an acre (mainly on chemicals, electricity, water and seed) to grow a crop that, in a good year, will earn him maybe $1,980. That's how much a french-fry processor will pay for the 20 tons of potatoes a single Idaho acre can yield. (The real money in agriculture—90 percent of the value added to the food we eat—is in selling inputs to farmers and then processing their crops.)

Danny Forsyth laid out the dismal economics of potato farming for me one sweltering morning at the coffee shop in downtown Jerome,

Idaho. Forsyth, 60, is a slight blue-eyed man with a small gray ponytail; he farms 3,000 acres of potatoes, corn and wheat, and he spoke about agricultural chemicals like a man desperate to kick a bad habit. "None of us would use them if we had any choice," he said glumly.

I asked him to walk me through a season's regimen. It typically begins early in the spring with a soil fumigant; to control nematodes, many potato farmers douse their fields with a chemical toxic enough to kill every trace of microbial life in the soil. Then, at planting, a systemic insecticide (like Thimet) is applied to the soil; this will be absorbed by the young seedlings and, for several weeks, will kill any insect that eats their leaves. After planting, Forsyth puts down an herbicide—Sencor or Eptam—to "clean" his field of all weeds. When the potato seedlings are six inches tall, an herbicide may be sprayed a second time to control weeds.

Idaho farmers like Forsyth farm in vast circles defined by the rotation of a pivot irrigation system, typically 135 acres to a circle; I'd seen them from 30,000 feet flying in, a grid of verdant green coins pressed into a desert of scrubby brown. Pesticides and fertilizers are simply added to the irrigation system, which on Forsyth's farm draws most of its water from the nearby Snake River. Along with their water, Forsyth's potatoes may receive 10 applications of chemical fertilizer during the growing season. Just before the rows close—when the leaves of one row of plants meet those of the next—he begins spraying Bravo, a fungicide, to control late blight, one of the biggest threats to the potato crop. (Late blight, which caused the Irish potato famine, is an airborne fungus that turns stored potatoes into rotting mush.) Blight is such a serious problem that the E.P.A. currently allows farmers to spray powerful fungicides that haven't passed the usual approval process. Forsyth's potatoes will receive eight applications of fungicide.

While I was in Idaho and Washington State, I asked potato farmers to show me their refuges. This proved to be a joke.

Twice each summer, Forsyth hires a crop duster to spray for aphids. Aphids are harmless in themselves, but they transmit the leafroll virus, which in Russet Burbank potatoes causes net necrosis, a brown spotting that will cause a processor to reject a whole crop. It happened to Forsyth [in 1997]. "I lost 80,000 bags"—they're a hundred pounds each—"to net necrosis," he said. "Instead of getting $4.95 a bag, I had to take $2 a bag from the dehydrator, and I was lucky to get that." Net necrosis is a purely cosmetic defect; yet because big buyers like McDonald's believe (with good reason) that we don't like to see brown spots in our fries, farmers like Danny Forsyth must spray their fields with some of the most toxic chemicals in use, including an organophosphate called Monitor.

"Monitor is a deadly chemical," Forsyth said. "I won't go into a field for four or five days after it's been sprayed—even to fix a broken pivot." That is, he would sooner lose a whole circle to drought than expose himself or an employee to Monitor, which has been found to cause neurological damage.

It's not hard to see why a farmer like Forsyth, struggling against tight margins and heartsick over chemicals, would leap at a New Leaf—or, in his case, a New Leaf Plus, which is protected from leafroll virus as well as beetles. "The New Leaf means I can skip a couple of sprayings, including the Monitor," he said. "I save money, and I sleep better. It also happens to be a nice-looking spud." The New Leafs don't come cheaply, however. They cost between $20 and $30 extra per acre in "technology fees" to Monsanto.

Forsyth and I discussed organic agriculture, about which he had the usual things to say ("That's all fine on a small scale, but they don't have to feed the world"), as well as a few things I'd never heard from a conventional farmer: "I like to eat organic food, and in fact I raise a lot of it at the house. The vegetables we buy at the market we just wash and wash and wash. I'm not sure I should be saying this, but I always plant a small area of potatoes without any chemicals. By the end of the season, my field potatoes are fine to eat, but any potatoes I pulled today are probably still full of systemics. I don't eat them."

Forsyth's words came back to me a few hours later, during lunch at the home of another potato farmer. Steve Young is a progressive and prosperous potato farmer—he calls himself an agribusinessman. In addition to his 10,000 acres—the picture window in his family room gazes out on 85 circles, all computer-controlled—Young owns a share in a successful fertilizer distributorship. His wife prepared a lavish feast for us, and after Dave, their 18-year-old, said grace, adding a special prayer for me (the Youngs are devout Mormons), she passed around a big bowl of home-made potato salad. As I helped myself, my Monsanto escort asked what was in the salad, flashing me a smile that suggested she might already know. "It's a combination of New Leafs and some of our regular Russets," our hostess said proudly. "Dug this very morning."

Refuges are illusive

After talking to farmers like Steve Young and Danny Forsyth, and walking fields made virtually sterile by a drenching season-long rain of chemicals, you could understand how Monsanto's New Leaf potato does indeed look like an environmental boon. Set against current practices, growing New Leafs represents a more sustainable way of potato farming. This advance must be weighed, of course, against everything we don't yet know about New Leafs—and a few things we do: like the problem of Bt resistance I had heard so much about back East. While I was in Idaho and Washington State, I asked potato farmers to show me their refuges. This proved to be a joke.

"I guess that's a refuge over there," one Washington farmer told me, pointing to a cornfield.

Monsanto's grower contract never mentions the word "refuge" and only requires that farmers plant no more than 80 percent of their fields in New Leaf. Basically, any field not planted in New Leaf is considered a refuge, even if that field has been sprayed to kill every bug in it. Farmers call such acreage a clean field; calling it a refuge is a stretch at best.

It probably shouldn't come as a big surprise that conventional farmers would have trouble embracing the notion of an insect refuge. To insist on real and substantial refuges is to ask them to start thinking of their fields in an entirely new way, less as a factory than as an ecosystem. In the factory, Bt is another in a long line of "silver bullets" that work for a while and then get replaced; in the ecosystem, all bugs are not necessarily bad, and the relationships between various species can be manipulated to achieve desired ends—like the long-term sustainability of Bt.

Growing organically

This is, of course, precisely the approach organic farmers have always taken to their fields, and after my lunch with the Youngs that afternoon, I paid a brief visit to an organic potato grower. Mike Heath is a rugged, laconic man in his mid-50's; like most of the organic farmers I've met, he looks as though he spends a lot more time out of doors than a conventional farmer, and he probably does: chemicals are, among other things, labor-saving devices. While we drove around his 500 acres in a battered old pickup, I asked him about biotechnology. He voiced many reservations—it was synthetic, there were too many unknowns—but his main objection to planting a biotech potato was simply that "it's not what my customers want."

That point was driven home [in] December [1997] when the Department of Agriculture proposed a new "organic standards" rule that, among other things, would have allowed biotech crops to carry an organic label. After receiving a flood of outraged cards and letters, the agency backed off. (As did Monsanto, which asked the U.S.D.A. to shelve the issue for three years.) Heath suggested that biotech may actually help organic farmers by driving worried consumers to the organic label.

I asked Heath about the New Leaf. He had no doubt resistance would come—"the bugs are always going to be smarter than we are"—and said it was unjust that Monsanto was profiting from the ruin of Bt, something he regarded as a "public good."

None of this particularly surprised me; what did was that Heath himself resorted to Bt sprays only once or twice in the last 10 years. I had assumed that organic farmers used Bt or other approved pesticides in much the same way conventional farmers use theirs, but as Heath showed me around his farm, I began to understand that organic farming was a lot more complicated than substituting good inputs for bad. Instead of buying many inputs at all, Heath relied on long and complex crop rotations to prevent a buildup of crop-specific pests—he has found, for example, that planting wheat after spuds "confuses" the potato beetles.

It was the very complexity of these fields—the sheer diversity of species, both in space and time—that made them productive year after year without many inputs.

He also plants strips of flowering crops on the margins of his potato fields—peas or alfalfa, usually—to attract the beneficial insects that eat beetle larvae and aphids. If there aren't enough beneficials to do the job, he'll introduce ladybugs. Heath also grows eight varieties of potatoes, on the theory that biodiversity in a field, as in the wild, is the best defense against any imbalances in the system. A bad year with one variety will probably be offset by a good year with the others.

"I can eat any potato in this field right now," he said, digging Yukon Golds for me to take home. "Most farmers can't eat their spuds out of the field. But you don't want to start talking about safe food in Idaho."

Heath's were the antithesis of "clean" fields, and, frankly, their weedy margins and overall patchiness made them much less pretty to look at. Yet it was the very complexity of these fields—the sheer diversity of species, both in space and time—that made them productive year after year without many inputs. The system provided for most of its needs.

All told, Heath's annual inputs consisted of natural fertilizers (compost and fish powder), ladybugs and a copper spray (for blight)—a few hundred dollars an acre. Of course, before you can compare Heath's operation with a conventional farm, you've got to add in the extra labor (lots of smaller crops means more work; organic fields must also be cultivated for weeds) and time—the typical organic rotation calls for potatoes every fifth year, in contrast to every third on a conventional farm. I asked Heath about his yields. To my astonishment, he was digging between 300 and 400 bags per acre—just as many as Danny Forsyth and only slightly fewer than Steve Young. Heath was also getting almost twice the price for his spuds: $8 a bag from an organic processor who was shipping frozen french fries to Japan.

Organic farming versus big business

On the drive back to Boise, I thought about why Heath's farm remained the exception, both in Idaho and elsewhere. Here was a genuinely new paradigm that seemed to work. But while it's true that organic agriculture is gaining ground (I met a big grower in Washington who had just added several organic circles), few of the mainstream farmers I met considered organic a "realistic" alternative. For one thing, it's expensive to convert: organic certifiers require a field to go without chemicals for three years before it can be called organic. For another, the U.S.D.A., which sets the course of American agriculture, has long been hostile to organic methods.

But I suspect the real reasons run deeper, and have more to do with the fact that in a dozen ways a farm like Heath's simply doesn't conform to the requirements of a corporate food chain. Heath's type of agriculture doesn't leave much room for the Monsantos of this world: organic farmers buy remarkably little—some seed, a few tons of compost, maybe a few gallons of ladybugs. That's because the organic farmer's focus is on a process, rather than on products. Nor is that process readily systematized, reduced to, say, a prescribed regime of sprayings like the one Forsyth outlined for me— regimes that are often designed by companies selling chemicals.

Most of the intelligence and local knowledge needed to run Mike Heath's farm resides in the head of Mike Heath. Growing potatoes conventionally requires intelligence, too, but a large portion of it resides in laboratories in distant places like St. Louis, where it is employed in developing sophisticated chemical inputs. That sort of centralization of agriculture is unlikely to be reversed, if only because there's so much money in it; besides, it's much easier for the farmer to buy prepackaged solutions from big companies. "Whose Head Is the Farmer Using? Whose Head Is Using the Farmer?" goes the title of a Wendell Berry essay.

Organic farmers like Heath have also rejected what is perhaps the cornerstone of industrial agriculture: the economies of scale that only a monoculture can achieve. Monoculture—growing vast fields of the same crop year after year—is probably the single most powerful simplification of modern agriculture. But monoculture is poorly fitted to the way nature seems to work. Very simply, a field of identical plants will be exquisitely vulnerable to insects, weeds and disease. Monoculture is at the root of virtually every problem that bedevils the modern farmer, and that virtually every input has been designed to solve.

To put the matter baldly, a farmer like Heath is working very hard to adjust his fields and his crops to the nature of nature, while farmers like Forsyth are working equally hard to adjust nature in their fields to the requirement of monoculture and, beyond that, to the needs of the industrial food chain. I remember asking Heath what he did about net necrosis, the bane of Forsyth's existence. "That's only really a problem with Russet Burbanks," he said. "So I plant other kinds." Forsyth can't do that. He's part of a food chain—at the far end of which stands a long, perfectly golden McDonald's fry—that demands he grow Russet Burbanks and little else.

The silver bullet myth

This is where biotechnology comes in, to the rescue of Forsyth's Russet Burbanks and, if Monsanto is right, to the whole food chain of which they form a part. Monoculture is in trouble—the pesticides that make it possible are rapidly being lost, either to resistance or to heightened concerns about their danger. Biotechnology is the new silver bullet that will save monoculture. But a new silver bullet is not a new paradigm—rather, it's something that will allow the old paradigm to survive. That paradigm will always construe the problem in Forsyth's fields as a Colorado potato beetle problem, rather than as a problem of potato monoculture.

Like the silver bullets that preceded them—the modern hybrids, the pesticides and the chemical fertilizers—the new biotech crops will probably, as advertised, increase yields. But equally important, they will also speed the process by which agriculture is being concentrated in a shrinking number of corporate hands. If that process has advanced more slowly in farming than in other sectors of the economy, it is only because nature herself—her complexity, diversity and sheer intractability in the face of our best efforts at control—has acted as a check on it. But biotechnology promises to remedy this "problem," too.

Controlling the seeds

Consider, for example, the seed, perhaps the ultimate "means of production" in any agriculture. It is only in the last few decades that farmers have begun buying their seed from big companies, and even today many farmers still save some seed every fall to replant in the spring. Brownbagging, as it is called, allows farmers to select strains particularly well adapted to their needs; since these seeds are often traded, the practice advances the state of the genetic art—indeed, has given us most of our crop plants. Seeds by their very nature don't lend themselves to commodification: they produce more of themselves ad infinitum (with the exception of certain modern hybrids), and for that reason the genetics of most major crop plants have traditionally been regarded as a common heritage. In the case of the potato, the genetics of most important varieties—the Burbanks, the Superiors, the Atlantics—have always been in the public domain. Before Monsanto released the New Leaf, there had never been a multinational seed corporation in the potato-seed business—there was no money in it.

Monsanto is using informants and hiring Pinkertons to enforce its patent rights.

Biotechnology changes all that. By adding a new gene or two to a Russet Burbank or Superior, Monsanto can now patent the improved variety. Legally, it has been possible to patent a plant for many years, but biologically, these patents have been almost impossible to enforce. Biotechnology partly solves that problem. A Monsanto agent can perform a simple test in my garden and prove that my plants are the company's intellectual property. The contract farmers sign with Monsanto allows company representatives to perform such tests in their fields at will. According to *Progressive Farmer*, a trade journal, Monsanto is using informants and hiring Pinkertons [Allan Pinkerton was a U.S. private detective (1819–1884)] to enforce its patent rights; it has already brought legal action against hundreds of farmers for patent infringement. . . .

At lunch on his farm in Idaho, I had asked Steve Young what he thought about all this, especially about the contract Monsanto made him sign. I wondered how the American farmer, the putative heir to a long tradition of agrarian independence, was adjusting to the idea of field men snooping around his farm, and patented seed he couldn't replant. Young said he had made his peace with corporate agriculture, and with biotechnology in particular: "It's here to stay. It's necessary if we're going to feed the world, and it's going to take us forward."

Then I asked him if he saw any downside to biotechnology, and he paused for what seemed a very long time. What he then said silenced the table. "There is a cost," he said. "It gives corporate America one more noose around my neck."

Harvesting the potatoes

A few weeks after I returned home from Idaho, I dug my New Leafs, harvesting a gorgeous-looking pile of white spuds, including some real lunkers. The plants had performed brilliantly, though so had all my other potatoes. The beetle problem never got serious, probably because the diversity of species in my (otherwise organic) garden had attracted enough beneficial insects to keep the beetles in check. By the time I harvested my crop, the question of eating the New Leafs was moot. Whatever I thought about the soundness of the process that had declared these potatoes safe didn't matter. Not just because I'd already had a few bites of New Leaf potato salad at the Youngs but also because Monsanto and the F.D.A. and the E.P.A. had long ago taken the decision of whether or not to eat a biotech potato out of my—out of all of our—hands. Chances are, I've eaten New Leafs already, at McDonald's or in a bag of Frito-Lay chips, though without a label there can be no way of knowing for sure.

So if I've probably eaten New Leafs already, why was it that I kept putting off eating mine? Maybe because it was August, and there were so many more-interesting fresh potatoes around—fingerlings with dense, luscious flesh, Yukon Golds that tasted as though they had been pre-buttered—that the idea of cooking with a bland commercial variety like the Superior seemed beside the point.

There was this, too: I had called Margaret Mellon at the Union of Concerned Scientists to ask her advice. Mellon is a molecular biologist and lawyer and a leading critic of biotech agriculture. She couldn't offer any hard scientific evidence that my New Leafs were unsafe, though she emphasized how little we know about the effects of Bt in the human diet. "That research simply hasn't been done," she said.

I pressed. Is there any reason I shouldn't eat these spuds?

"Let me turn that around. Why would you want to?"

It was a good question. So for a while I kept my New Leafs in a bag on the porch. Then I took the bag with me on vacation, thinking maybe I'd sample them there, but the bag came home untouched.

The bag sat on my porch till the other day, when I was invited to an end-of-summer potluck supper at the town beach. Perfect. I signed up to make a potato salad. I brought the bag into the kitchen and set a pot of water on the stove. But before it boiled I was stricken by this thought: I'd have to tell people at the picnic what they were eating. I'm sure (well, al-most sure) the potatoes are safe, but if the idea of eating biotech food without knowing it bothered me, how could I possibly ask my neighbors to? So I'd tell them about the New Leafs—and then, no doubt, lug home a big bowl of untouched potato salad. For surely there would be other potato salads at the potluck and who, given the choice, was ever going to opt for the bowl with the biotech spuds?

So there they sit, a bag of biotech spuds on my porch. I'm sure they're absolutely fine. I pass the bag every day, thinking I really should try one, but I'm beginning to think that what I like best about these particular biotech potatoes—what makes them different—is that I have this choice. And until I know more, I choose not.

3

Genetically Engineered Foods Have Health and Environmental Benefits

Institute of Food Technologists

The Institute of Food Technologists (IFT) is a nonprofit society for food science and technology. The IFT conducted a review to evaluate the current scientific information concerning biotechnology. The following information is taken from the review written by three panels of experts consisting of IFT members and other biotechnology authorities and was published in August, September, and October of 2000 in Food Technology.

Among the specific benefits of recombinant DNA (rDNA) technology, also know as genetic engineering or modification of foods, is the development of disease resistance in crops such as pears and apples. Biotechnology has produced the first trees resistant to devastating "fire blight" bacterial disease. Another benefit is the production of foods with improved nutritional value, including increased quantifies of essential amino acids, vitamins, and necessary fatty acids. Biotechnology can enhance foods for millions of poor children with specific nutritional deficiences. Other possible benefits include decreasing natural allergenic proteins in certain foods and producing edible vaccines for preventing diseases. Numerous national and international scientific groups have reached the conclusion that rDNA food is safe to eat. Farmers and producers have embraced rDNA biotech crops due to a decreased need to apply environmentally harmful herbicides and pesticides.

New technologies rarely receive a broad and enthusiastic welcome. Canned food, for its first hundred years, was viewed apprehensively, and not without reason. In those pre-bacteriology days, it was far more an uncertain art than a solid science. Pasteurized milk, a life-saving technology in its elimination of the microorganisms that cause tuberculosis and undulant fever, was originally viewed with deep suspicion. Artificial insemination of farm animals—critical in selective breeding of improved livestock—was regarded as tampering with nature. Recombinant DNA biotechnology is no exception.

Although rDNA biotechnology offers numerous benefits, it also has raised several issues of consumer concern. A thorough analysis of these concerns reveals that many stem from not fully understanding the science involved and how these potential risks have already been addressed. This report explores the benefits and evaluates the concerns in order to contribute to civil and rational dialogue which alone can deal effectively with both scientific issues and consumer concerns about this new technology.

Specific benefits

Stronger Plants. Recombinant DNA biotechnology is the latest in a long line of tools that plant breeders have used to enhance plant availability, survival, and growth to benefit people. In little more than a century, starting with hybridization, which was commercialized in the first decade of the 20th century, scientific breakthroughs enabled new types of plants, such as seedless watermelons and grapes, to be produced.

Plant breeding often has been successful in producing plants with increased pest and disease resistance, while retaining high yields, taste, and processing attributes. Apple and pear production is constrained by the bacterial disease called fire-blight, first described in the 1870s. No satisfactory antibacterial compounds or adequate resistance to the disease is available in apples desired by consumers. Recombinant DNA biotechnology research has produced the first trees to resist this devastating disease.

The susceptibility of a plant to biotic and environmental stresses, such as temperature extremes, exposure to heavy metals such as aluminum, and salt and drought tolerance is heavily affected by the plant's genetic composition and structure. For example, some leaves have evolved to conserve moisture and resist heat or freezing. Breeders have changed leaf and stem architecture to capture more sunlight and allow for greater air flow through the leaf canopy.

Improved Nutrition. Specific foods can be developed to correct malnutrition problems that are unique to different regions of the world. To this end, plants can be modified to provide increased and more stable quantities of essential amino acids, vitamins, or desirable fatty acids. For example, "golden rice" has been genetically modified through rDNA biotechnology to have increased betacarotene content, which may help to overcome the severe vitamin A deficiencies which cause millions of poor children to go blind or die every year in low-income, rice-consuming cultures. A related product of rDNA biotechnology may also help eliminate the iron deficiency that threatens hundreds of millions of women and babies with birth complications each year.

Probiotics are living microorganisms, typically delivered through foods, that offer benefits to health and well being that are beyond basic nutrition. Selected members within the *Lactobacillus* and *Bifidobacterium* genera are considered key probiotic species because they survive passage through the gastrointestinal tract and exert benefits there; such as stimulation of the immune system and balancing a healthy microbial flora. Recombinant DNA biotechnology is expected to play an important role in identifying the probiotic strains capable of eliciting certain health benefits.

Higher Crop Yields. Plants also can be modified to grow well in areas of low production potential. For example, toxic metals, such as aluminum and manganese, are widely present in "acidic" tropical soils, which account for nearly half the arable land in the tropics. These metals reduce root growth, cutting yields by up to 80 percent. To produce acid-tolerant crops, two researchers in Mexico inserted a gene from a bacterium into tobacco and papaya. The plants thus secrete citric acid from their roots, chelating these toxic metals. The yield gains now anticipated from making such soils accessible will be critical to protecting the tropical forests, which contain most of the world's species of plants and animals.

Reduced Allergenicity. Recombinant DNA biotechnology also offers the opportunity to decrease or eliminate the allergenic proteins that occur naturally in specific foods. For example, rDNA biotechnology has already been used to dramatically reduce the levels of the major rice allergen. Similar approaches could be attempted with more commonly allergenic foods such as peanuts.

Medical Benefits. Recombinant DNA biotechnology brings closer to reality the prospect of commercial production in plants of edible vaccines and therapeutics for preventing and treating animal and human diseases. Possibilities include a wide variety of compounds, ranging from vaccine antigens for hepatitis B and Norwalk viruses, bacteria *(Pseudomonas aeruginosa and Staphylococcus aureus)*, to vaccines against cancer and diabetes. In addition, genetically-modified strains of probiotic microorganisms are also possible vehicles for successful delivery of vaccines and digestive aids, such as lactase, through the stomach and the small intestine. Recombinant DNA biotechnology-derived vaccines are potentially cheap, convenient to distribute, and simple and safe to administer. In 1998, scientists reported the first successful human clinical trials with an edible vaccine against a pathogenic strain of *Escherichia coli.*

Specific foods can be developed to correct malnutrition problems that are unique to different regions of the world.

Researchers also have developed a system to produce in tobacco plants a therapeutic vaccine against non-Hodgkin's B-cell lymphoma in mice. Eighty percent of mice receiving the plant-derived vaccine survived the lymphoma, while all untreated mice died within three weeks after contracting the disease. A similar approach was used to develop a vaccine against insulin-dependent diabetes mellitus (IDDM). Insulin and pancreatic glutamic acid decarboxylase (GAD), linked to the onset of IDDM, are candidates for use as oral vaccines. Scientists have developed a potato-based insulin vaccine that is almost 100 times more powerful than the existing vaccine in preventing IDDM. Diabetes-prone mice fed potatoes engineered to produce GAD had reduced incidence of disease and immune response severity.

Healthier Farm Animals. Advances in genetics and rDNA biotechnology make it possible to envision ways of improving the nutritional content of animal feed by directing the plant to produce a more nutritious product. Plant breeders used rDNA biotechnology to develop a corn that is easier for farm animals to digest. A further improvement came with the development of nutritionally dense corn, which has increased amounts of oil, protein, and essential amino acids necessary for optimal animal growth. In all crops it seems reasonable to expect additional improvement through further enhancements in oils and fatty acids, protein, starch, and carbohydrates, and enhancements in vitamins, antioxidants, and mineral composition. Recombinant DNA biotechnology has also helped develop an animal feed corn with lower levels of phytate. This improvement reduces phosphorous and nitrogen in animal waste. People who live near large farms especially benefit from this improvement, because it reduces the intense odors that may waft their way.

Farmers and producers have enthusiastically embraced the new varieties of rDNA biotechnology-derived crops that exhibit increased resistance to insects.

The techniques being used to develop edible human vaccines are also being applied to develop vaccines for animals. Researchers have already demonstrated production in plants of a vaccine against transmissible gastroenteritis virus, which protected swine in clinical trails against the virulent pathogen.

Environmental Benefits. Farmers and producers have enthusiastically embraced the new varieties of rDNA biotechnology-derived crops that exhibit increased resistance to insects. Examples include corn and cotton with *Bacillus thuringiensis* (Bt) genes for insecticidal proteins, tolerance to herbicides (corn, cotton, soybeans), and virus-resistant crops including squash, cucumbers, and papaya. In 1998, 45 percent of farmers had higher yields of Bt corn compared to conventional corn, and nearly 26 percent of farmers growing Bt corn reported a decrease in pesticide use.

Recombinant DNA biotechnology also makes it possible to use herbicides that are less harmful to the environment. The herbicide glyphosate, for example, rapidly degrades in soil, breaking down into harmless carbon dioxide and water, unlike earlier herbicides that persisted in the environment and contaminated groundwater. Glyphosate is replacing these herbicides with the introduction of rDNA biotechnology-derived, glyphosate-tolerant plant crops such as canola and soybeans. As an added benefit, these canola plants require only one herbicide application, instead of two.

New Ingredients. Recombinant rDNA biotechnology can lead to improvements in microorganisms such as bacteria, yeasts, and molds that help convert milk, cereals, vegetables, and meats into a plethora of fermented products, including cheese, cultured milk, sourdough bread, pickles, and sausages. It also can lead to improvements in production of basic ingredients and nutrients such as organic acids, amino acids, vitamins and enzymes.

Enzymes have historically played a key role in production of bread, cheese, and alcoholic beverage production. Today, enzymes remain indispensable to modern food processing, and many are produced using rDNA biotechnology. Chymosin, used to clot milk in cheesemaking, was the first enzyme produced by rDNA biotechnology for use in food. Traditionally, chymosin was obtained from rennet extracted from the fourth stomach of young calves. Rennet supplies faced major declines as calf slaughter decreased during a period of increasing worldwide cheese production. Chymosin produced through rDNA biotechnology is substantially more pure than traditional rennet. The Food and Drug Administration (FDA) concluded in 1990 that rDNA biotechnology-derived chymosin is identical to its natural counterpart and, therefore, is acceptable for use in foods. Use of rDNA biotechnology-derived chymosin now exceeds 80 percent of the global market.

Food Safety Improvements. Preliminary studies have shown that rDNA biotechnology-derived foods and food ingredients may have food safety benefits. For example, preliminary studies by the U.S. Department of Agriculture's (USDA) Agricultural Research Service have shown that Bt corn had lower levels of fumonisin, a potential cancer-causing agent often found at elevated levels in insect-damaged kernels. In Bt corn, fumonisin levels were 30-to-40 fold lower than in non-Bt corn varieties. Mycotoxins like fumonisin are both a public health issue and an export issue, as European and Asian markets have refused to import U.S. corn due to unacceptable levels of mycotoxins. It may be possible to create corn varieties with greater resistance to a variety of insects, leading to lower levels of mycotoxin contamination.

Economic Benefits. The most widespread rDNA biotechnology-derived crops in the United States are soybean, cotton and corn. In 1999, 35 percent of U.S. corn acreage (77.4 million acres) was made up of either insect-tolerant (23 percent) or herbicide-tolerant cultivars; 45 percent of the cotton acreage (14.8 million acres) was insect tolerant; and 54 percent of the soybean acreage (72.9 million acres) was herbicide tolerant. A 1997 USDA study found that herbicide-tolerant soybeans reduced farm input costs by 3 to 6 percent and increased average yields by more than 13 to 18 percent in most regions of the United States. . . .

Frequently asked questions about human food safety

Is rDNA biotechnology-derived food safe to eat?

The bottom line answer is yes, rDNA biotechnology-derived food is safe to eat. In fact, biotechnology processes are more precise and predictable than conventional cross breeding techniques. Transferring one or two well characterized genes through rDNA biotechnology techniques presents less risk of unexpected consequences than conventional cross breeding, where all 100,000 or so of each plant's genes mix and result in random combinations. Risk-based safety assessment of rDNA biotechnology-derived foods compares the new plant variety to its traditional counterpart to identify and evaluate any differences that might present safety or nutritional concerns. Numerous national and international scientific groups have concluded rDNA biotechnology-derived foods are safe.

Will eating rDNA biotechnology-derived foods alter my DNA?

No. DNA breaks down readily and completely during digestion.

Are consumers eating large quantities of rDNA?

No. The DNA introduced using rDNA biotechnology represents only 1/250,000 of the total DNA consumed when the food is eaten. There is no evidence for any transfer of DNA from conventional foods, and virtually no chance it could ever occur. The risk of transfer of rDNA segments is far less than for DNA from conventional foods.

Have any people become ill because they ate rDNA biotechnology-derived food?

No. No adverse effects have been reported. Inaccurate claims of illness and death from consumption of rDNA biotechnology-derived food are based on an incident in the late 1980s in which consumption of L-tryptophan produced by an rDNA biotechnology-derived organism caused the illness of 1,500 people and the death of 37 in the United States. However, overwhelming evidence has proven that the illness was caused by the Japanese manufacturer's failure to perform standard purification to remove impurities. Careful purification is a necessary procedure when a microorganism is used in food ingredient production, as well as other food production methods. Because the manufacturer of the L-tryptophan began using the rDNA biotechnology-derived organism at the same time it eliminated reverse osmosis and reduced the amount of activated carbon in the purification process, some observers incorrectly attributed the incident to the use of biotechnology.

Can scientists prove that rDNA biotechnology-derived foods are safe to eat?

Unfortunately, it is impossible to prove the absolute safety of any food. Science just doesn't work that way. Scientists use the vast amount of historical knowledge about the safety of various foods to assess the risks posed by a new food. Comparing a new plant variety to its traditional counterpart highlights potential differences that might present safety or nutritional concerns. If a substance in a new food has been safely consumed in another food, it is generally considered safe. In contrast, the introduction of a substance that is completely new to the food supply triggers further study to ensure it does not contain a toxin, an antinutrient, or an allergen. Focusing on the characteristics that are different from the traditional counterpart is an effective, risk-based approach that has been thoroughly considered and widely employed by numerous national and international scientific organizations.

The impressive safety record for both conventional and rDNA biotechnology-derived foods demonstrates the effectiveness of the safety assessment.

Preliminary studies have shown that rDNA biotechnology-derived foods and food ingredients may have food safety benefits.

Have rDNA biotechnology-derived foods been tested for safety?

Yes. All the rDNA biotechnology-derived foods on the market to date have undergone a science-based safety assessment. All foods on the market, rDNA biotechnology-derived or conventional, are subject to FDA's high standards for food safety.

What risks does the current risk-based safety assessment consider?

The safety assessment considers all aspects of food safety, including microbiological issues, toxicity, allergenicity, and nutrition. In evaluating the safety of the DNA to be introduced, scientists consider whether the source of the gene(s) has a history of safe consumption and whether the source produces any toxins or allergens that would need to be assessed in the rDNA biotechnology-derived plant. Researchers also evaluate the proteins encoded by the introduced gene and other proteins that may be produced or altered by the presence of the introduced protein in the plant. The rDNA biotechnology-derived food is compared with its conventional counterpart to identify characteristics that require additional assessment for safety and nutritional issues.

Numerous national and international scientific groups have concluded rDNA biotechnology-derived foods are safe.

Why not conduct long-term feeding studies on rDNA biotechnology-derived foods before they are introduced into the food supply?

Animal feeding studies work relatively well for the safety testing of minor food components—food additives, naturally occurring trace constituents, pesticide residues, etc. They do not work well for major ingredients and whole foods. It simply is not possible to cram enough of the food into an animal's diet. Any effort to do so typically distorts the animal's diet so much that severe nutritional problems and, occasionally, the toxicity of naturally occurring constituents, contaminate the study results. Frequently the study must be discontinued. Moreover, all this happens at feeding levels that are still far short of those required to demonstrate an adequate margin of safety. Thus, it is far more productive to focus safety studies on any aspects of rDNA biotechnology-derived foods that are different from their conventional counterparts.

Will rDNA biotechnology-derived food cause allergic reactions?

No adverse effects have been reported. Evaluating the potential allergenicity of the introduced protein is an important step in the safety assessment process. Virtually all food allergens are proteins, although only a small fraction of the proteins found in nature are allergenic. The assessment of potential allergenicity considers whether the gene is from a source that is known to cause allergies, the chemical similarity of the introduced protein to known allergens, and other characteristics such as the digestive stability of the protein. As scientists learn more about the mechanisms of food allergies in general, the criteria used to assess the allergenicity of rDNA biotechnology-derived foods will continue to be refined.

What if the new DNA produces unexpected changes in the food, like increased levels of toxins?

There is no scientific evidence of the occurrence of unexpected toxic or antinutrient factors or of enhanced production of known toxic constituents in any rDNA biotechnology-derived foods brought to market. In the event such unexpected changes did occur, they would be identified as part of the safety assessment and subject to additional study. The more precise and predictable nature of genetic change accomplished using rDNA techniques means such unexpected effects would be less likely in foods derived from rDNA biotechnology than in foods produced using conventional breeding techniques, but the testing is still conducted.

If rDNA biotechnology-derived foods are safe, why are some people worried about them?

That is a difficult question to answer because people form their initial impressions of new products and new technologies based on many different factors, of which science and safety are only two. Cost, convenience, familiarity, esthetics, and "what's in it for me?" are some of the others. Even life-saving new medical technologies have often required years for full acceptance. In the case of biotechnology, many of the specific safety concerns some consumers have expressed are not supported by the scientific evidence. Without a thorough understanding of rDNA biotechnology, it is difficult to understand what the real potential risks are and how they are addressed. One purpose of the IFT Expert Report on Biotechnology and Foods was to provide a comprehensive review of the science to explain how safety is assured and the basis for conclusions of safety.

4

Genetically Engineered Foods Have Health Risks

Roberto Verzola

Roberto Verzola is the secretary-general of the Philippine Greens, a group of political activists advocating the principles of ecology, social justice, and self-determination.

Researchers have discovered that genetically engineered (GE) foods can have negative health effects. These foods can unleash new pathogens, contain allergens and toxins, and increase the risk of cancer, herbicide exposure, and harm to fetuses and infants. Because the biotech industry has represented GE foods as "substantially equivalent" to conventional crops, and because a sympathetic Food and Drug Administration has supported the industry, essential tests on GE foods have not been done. Public resistance in the face of these dangers has caused some biotech companies to cut back on their GE programs.

The year 1999 marks what might be the turning point in the global fight against genetically-engineered (GE) food, as the issue began to grip media and public attention. Renewed debate flared in the British media in February 1999, when some 20 scientists from 13 countries issued a statement deploring the harsh treatment by Scotland's Rowett Research Institute of world-renowned British researcher and lectin expert Dr. Arpad Pusztai and demanding his reinstatement. Pusztai had earlier begun a £1.6-million study which indicated that a GE potato diet weakened rats' immune systems and adversely affected the animals' internal organs. When he shared with the media (with his superior's permission) some of his concerns, Pusztai was promptly sacked from his research post. His papers were confiscated, he was prohibited from talking to the media, and his research team was closed down.

Pusztai case prods anti-GE backlash

The strong statement by the 20 scientists calling for a review of Pusztai's case launched a wave of investigative reports, bringing into the open many of the little-known unresolved questions about GE food safety. Surveys revealed increasing consumer aversion to GE food. Market response was swift. Reacting to clearly-expressed consumer preferences, one food processor and distributor after another announced that they were keeping their products and shelves GE-free.

The British Medical Association (BMA), which counts some 80% or nearly 115,000 of Britain's medical doctors, issued an official statement expressing concern over the safety of GE foods.

In May 1999, the British Medical Association (BMA), which counts some 80% or nearly 115,000 of Britain's medical doctors, issued an official statement expressing concern over the safety of GE foods. The BMA recommended a moratorium on planting commercial GE crops in the UK "until there is scientific consensus (or as close agreement as reasonably achievable) about the potential long-term environmental effects." The BMA also called for (1) segregation at source, "to enable identification and traceability" of GE foods; (2) labeling GE imports and banning unlabelled ones if the industry refuses to segregate; and (3) more robust systems of disease surveillance, to deal with "potential emergence of new diseases associated with genetically modified (GM) material which will be obscure and difficult to diagnose."

Also in May, Cornell University assistant professor John Losey and his colleagues announced a study, which showed that Bacillus thuringiensis (Bt) corn pollen was deadly to monarch butterflies. Within days, the bad news made the headlines, bringing into the US public's attention the unresolved issues about GE food safety.

The monarch study launched another anti-GE wave. Europe banned the importation of Bt corn, and major US grain traders like Archer Daniels Midland announced they were only buying corn that Europe would accept. Baby food manufacturer Gerber promised not only to keep their products GE-free, but also to shift to organic ingredients. That says a lot, considering that Gerber is owned by Swiss giant and GE seed producer Novartis.

Aside from the US and a handful of others, most countries have adopted or are now contemplating mandatory labeling for GE food. And an increasing number of firms are committing to GE-free food products.

Problems with GE food

In the past, the most common examples cited against GE food safety had been: (1) the L-tryphtophan case (where a toxic contaminant that was likely due to the GE bacteria used in producing the food supplement L-tryphtophan killed 37 and hospitalized 1,500); (2) the Brazil nut case (where soya inserted with a Brazil nut gene to raise its protein content also showed increased allergenic effects); and (3) the rBGH case (where the use of a GE growth hormone to stimulate milk production in cows resulted in udder inflammations, infections and other problems affecting milk quality).

The year 1999 presented a rich set of new data justifying health and safety concerns. Some examples:

Allergenicity: Researchers of the York Nutritional Laboratory (UK) reported that health complaints in the UK involving soya—the leading GE food—have increased 50% (from 10 to 15 per 100 patients) in 1998. For the first time in 17 years of testing, soya joined the lab's top 10 allergy-causing foods.

Toxicity: Dr. Arpad Pusztai found that a diet of potatoes engineered to express the snowdrop lectin weakened rats' immune systems and adversely affected the kidney, thymus, spleen, gut and brain of the animals. If confirmed, Pusztai's conclusions will reinforce concerns that gene insertion itself may create new toxins; it will also implicate the toxin commonly used in other GE crops—the Bt toxin which, Pusztai says, is also a lectin.

The emergence of new pathogens: Mae-Wan Ho and Angela Ryan of the UK Open University warned last July 1999 that "no transgenic plant containing the CaMV promoter should be released," because the Cauliflower Mosaic Virus (CaMV) promoter is "very likely to recombine with other DNA in the host genome, including dormant viral DNA, as well as with other viruses in the host cell." The problem covers practically all GE plants released so far. These GE plants, according to Ryan, "have the potential to create new viruses or other invasive genetic elements." Earlier, the British Medical Association had also called for a "ban on the use of antibiotic resistance marker genes in GM food, as the risk to human health from antibiotic resistance developing in micro-organisms is one of the major public health threats that will be faced in the 21st Century." Norway has already banned the use of these marker genes; Europe is also considering a ban.

Aside from the US and a handful of others, most countries have adopted or are now contemplating mandatory labeling for GE food.

The risk of cancers: US food campaigner Robert Cohen warns about the hormone Insulin-like Growth Factor-1 (IGF-1), identical versions of which occur in cows and humans. In 1994, Cohen says, the US FDA approved the use of a GE hormone (rBGH) in cows to stimulate milk production. Using rBGH raises IGF-1 levels in cows' milk by 80%. IGF-1, Cohen warns, is a key factor in prostate cancer, breast cancer, and lung cancer. Most recently, Cohen cites a report in the Journal of the American Dietetic Association, which found IGF-1 levels in the blood of milk drinkers 10% higher than in non-drinkers. The implication: GE milk exposes its drinkers to higher cancer risks.

Higher herbicide exposures: Since herbicide-resistant GE crops lead to greater herbicide use, cancer risk can also come from exposure to higher levels of herbicides like bromoxynil (Rhone-Poulenc's Buctril) and glyphosate (Monsanto's Roundup). Authors Marc Lappe and Britt Bailey warn that bromoxynil bioaccumulates, because it is fat-soluble. Rat and rabbit studies have shown birth defects, other developmental disorders in fetuses, tumors, and carcinomas at levels ranging from 20 to 300 parts per million. Glyphosate exposure, on the other hand, can triple the risk of non-Hodgkin's lymphoma, say cancer specialists Dr. Lennart Hardell and Dr. Mikael Eriksson of Sweden's Orebro Hospital, in a study published in the American Cancer Society's journal.

By testing to show "substantial equivalence," biotech firms convinced a sympathetic US FDA to approve commercialization, and avoided actual feeding tests.

Higher risks for fetuses and babies: Lappe and Bailey also noted the "remarkably high estrogenic activity of soy isoflavones," elevated levels of which have been found in herbicide-treated GE soya. "If ingested by nursing infants, these isoflavones can produce circulating levels equivalent to 13,000 to 22,000 times the normal plasma estradiol concentrations found in babies, with unknown and potentially dangerous secondary effects," they warned. Early exposure to estrogens, they wrote, is associated with sex organ dysfunctions and higher risks of vaginal adenocarcinoma and other tumors. The concern of pediatric neurologist Dr. Martha Herbert of the Council for Responsible Genetics is "the immature gut and immature body of infants." If introduced too early, even proteins that are normally part of our diet can lead to auto-immune and allergic reactions later on, she said. "If a substance harms adults, it will harm babies, the sick and the elderly more severely, and after smaller exposures," Dr. Herbert warned in her June 1999 statement.

Inadequate safety studies: When Pusztai began his GE potato research in 1996, only one feeding test had been published—done by Monsanto (no harmful effects observed). A second feeding study on broiler chickens by a Novartis researcher was published 1998 (no harmful effects observed). Pusztai's rat study published October 1999—the only independent study so far—observed some harmful effects. No feeding studies had been done on swine or cattle (major consumers of GE corn and soya) or human volunteers. No study on the long-term effects of GE food had been done either. And if research on these novel toxins is just beginning, studies of their effects in combination with other toxins are nearly non-existent.

Why GE foods got marketed

If safety studies are lacking and scientific consensus is absent, how did GE foods get approved so quickly? Industry used a tricky argument, which goes this way: Genetic engineering is basically the same as conventional breeding. Thus, GE foods are "substantially equivalent" to their conventional counterparts, and no special tests should be required. By testing to show "substantial equivalence," biotech firms convinced a sympathetic US FDA to approve commercialization, and avoided actual feeding tests.

In September 1996, the World Health Organization (WHO) and UN Food and Agricultural Organization (FAO) convened an "expert" consultation on GE food safety in Rome, which adopted the same industry line that: (1) safety issues in GE foods were "basically of the same nature" as in foods from conventional breeding; (2) the substantial equivalence concept can be used to show GE food safety; and (3) once substantial equivalence is shown, "no further safety consideration is needed." While the meeting's report included a disclaimer that the participants were invited "in their individual capacities and not as representative of any organization, affiliation or government," biotech firms keep referring to this 1996 report to falsely claim that the "WHO/FAO have declared that Bt corn [or some other GE product] is as safe as its conventional equivalent for animal and human consumption." Yet the WHO and the FAO themselves have no such official position.

Today, scientists themselves question substantial equivalence as "a commercial and political judgment masquerading as if it were scientific . . . primarily to provide an excuse for not requiring biochemical or toxicological tests." (Letter to *Nature* by Erik Millstone, Eric Brunner & Sue Mayer, 10/7/99). The Codex Alimentarius, to which WHO and FAO defer on food safety issues, has not adopted the concept for its food safety assessments.

May be turning point for GE food industry

A final indication that marks 1999 as the turning point was supplied not by activists or biotech critics but by the industry itself:

"Ag Biotech: Thanks, But No Thanks?"—That was the title of a July 1999 report of investment analysts Frank Mitsch and Jennifer Mitchell of the Deutsche Banc Alex. Brown, the largest investment firm in the world. The two said they were "willing to believe that genetically modified organism (GMO) crops are safe," but they warned that the "no thanks" attitude "appears to be in the lead in Europe and could easily become the thought process in the United States as well." Earlier, three analysts from the same company had sent investors a report entitled "GMOs Are Dead," advising them to sell their Pioneer HiBred stock.

Terminator seeds: Monsanto announced in October 1999 that it was dropping its Terminator seed program, confirming the effectiveness of the global campaign against the technology of developing sterile GE seeds.

The monster is hurt; the challenge to GE food critics in the year 2000 and after is to find the jugular and go for it.

5

Genetic Engineering Threatens Biodiversity

Barbara Kingsolver

Barbara Kingsolver is a well-known author. She has worked as a biological researcher, an environmental activist, and a science writer with articles published in the Nation, *the* New York Times, *and* Smithsonian.

Genetic diversity is nature's way of assuring that species will endure over time, weathering all types of environmental changes. This diversity has been compromised in modern agriculture because a few large corporations control agriculture and sell relatively few varieties of seeds, resulting in crops that are genetically uniform. When this shallow gene bank is threatened, as with widespread plant diseases, researchers must return to the more diverse original strains of the "land races" grown by farmers mostly in the poorer parts of the world. With the introduction of new genetically engineered crops, the old seeds of the land races die out, cancelling nature's insurance policy. No crops could be engineered that would have the resilience of the old seeds, for genetically engineered genes do not have the inherent survival capability of genes that have evolved over three billion years.

At the root of everything, Darwin said, is that wonder of wonders, genetic diversity. You're unlike your sister, a litter of pups is its own small Rainbow Coalition, and every grain of wheat in a field holds inside its germ a slightly separate destiny. You can't see the differences until you cast the seeds on the ground and grow them out, but sure enough, some will grow into taller plants and some shorter, some tougher, some sweeter. In a good year all or most of them will thrive and give you wheat. But in a bad year a spate of high winds may take down the tallest stalks and leave standing at harvest time only, say, the 10 percent of the crop that had a "shortness" gene. And if that wheat comprises your winter's supply of bread, plus the only seed you'll have for next year's crop, then you'll be almighty glad to have that small, short harvest. Genetic diversity, in domestic populations as well as wild ones, is nature's sole insur-

43

ance policy. Environments change: Wet years are followed by droughts, lakes dry up, volcanoes rumble, ice ages dawn. It's a big, bad world out there for a little strand of DNA. But a population will persist over time if, deep within the scattered genetics of its ranks, it is literally prepared for anything. When the windy years persist for a decade, the wheat population will be overtaken by a preponderance of shortness, but if the crop maintains its diversity, there will always be recessive aspirations for height hiding in there somewhere, waiting to have their day.

We still rely on the gigantic insurance policy provided by the genetic variability in the land races, which continue to be hand-sown and harvested, year in and year out.

How is the diversity maintained? That old black magic called sex. Every seed has two parents. Plants throw their sex to the wind, to a hummingbird's tongue, to the knees of a bee—in April you are *inhaling* sex, and sneezing—and in the process, each two parents put their scrambled genes into offspring that represent whole new genetic combinations never before seen on Earth. Every new outfit will be ready for *something*, and together—in a large enough population—the whole crowd will be ready for *anything*. Individuals will die, not at random but because of some fatal misfit between what an organism *has* and what's *required*. But the population will live on, moving always in the direction of fitness (however "fitness" is at the moment defined), not because anyone has a master plan but simply because survival carries fitness forward, and death doesn't.

People have railed at this reality, left and right, since the evening when a British ambassador's wife declared to her husband, "Oh dear, let us hope Mr. Darwin isn't right, and if he is, let us hope no one finds out about it!" Fundamentalist Christians seem disturbed by a scenario in which individual will is so irrelevant. They might be surprised to learn that Stalin tried to ban the study of genetics and evolution in Soviet universities for the opposite reason, attacking the idea of natural selection—which acts only at the level of the individual—for being anti-Communist. Through it all, the little engines of evolution have kept on turning as they have done for millennia, delivering us here and passing on, untouched by politics or what anybody thinks.

Agriculture arises with diversity

Nikolai Vavilov was an astounding man of science, and probably the greatest plant explorer who has ever lived. He spoke seven languages and could recite books by [Russian poet Aleksandr] Pushkin from memory. In his travels through sixty-four countries between 1916 and 1940, he saw more crop diversity than anyone had known existed, and founded the world's largest seed collection.

As he combed continents looking for primitive crop varieties, Vavilov noticed a pattern: Genetic variation was not evenly distributed. In a small region of Ethiopia he found hundreds of kinds of ancient wheat known only to that place. A single New World plateau is astonishingly rich in corn varieties, while another one is rolling in different kinds of potatoes. Vavilov mapped the distribution of what he found and theorized that the degree of diversity of a crop indicated how long it had been grown in a given region, as farmers saved their seeds through hundreds and thousands of seasons. They also saved more *types* of seed for different benefits; thus popcorn, tortilla corn, roasting corn, and varieties of corn with particular colors and textures were all derived, over centuries, from one original strain. Within each crop type, the generations of selection would also yield a breadth of resistance to all types of pest and weather problems encountered through the years. By looking through his lens of genetics, Vavilov began to pinpoint the places in the world where human agriculture had originated. More modern genetic research has largely borne out his hypothesis that agriculture emerged independently in the places where the most diverse and ancient crop types, known as land races, are to be found: in the Near East, northern China, Mesoamerica, and Ethiopia.

The industrialized world depends entirely on crops and cultivation practices imported from what we now call the Third World (though evidently it was actually First). In an important departure from older traditions, the crops we now grow in the United States are extremely uniform genetically, due to the fact that our agriculture is controlled primarily by a few large agricultural corporations that sell relatively few varieties of seeds. Those who know the seed business are well aware that our shallow gene bank is highly vulnerable; when a crop strain succumbs all at once to a new disease, all across the country (as happened with our corn in 1970), researchers must return to the more diverse original strains for help. So we still rely on the gigantic insurance policy provided by the genetic variability in the land races, which continue to be hand-sown and harvested, year in and year out, by farmers in those mostly poor places from which our crops arose.

Unbelievably, we are now engaged in a serious effort to cancel that insurance policy.

A tragic story

It happens like this. Let's say you are an Ethiopian farmer growing a land race of wheat—a wildly variable, husky mongrel crop that has been in your family for hundreds of years. You always lose some to wind and weather, but the rest still comes through every year. Lately, though, you've been hearing about a kind of Magic Wheat that grows six times bigger than your crop, is easier to harvest, and contains vitamins that aren't found in ordinary wheat. And amazingly enough, by special arrangement with the government, it's free.

Readers who have even the slightest acquaintance with fairy tales will already know there is trouble ahead in this story. The Magic Wheat grows well the first year, but its rapid, overly green growth attracts a startling number of pests. You see insects on this crop that never ate wheat before, in the whole of your family's history. You watch, you worry. You realize that you're going to have to spray a pesticide to get this crop through to harvest. You're not so surprised to learn that by special arrangement with the government, the same company that gave you the seed for free can sell you the pesticide you need. It's a good pesticide, they use it all the time in America, but it costs money you don't have, so you'll have to borrow against next year's crop.

If genetically reordered organisms escape into natural populations, they may rapidly change the genetics of an entire species in a way that could seal its doom.

The second year, you will be visited by a terrible drought, and your crop will not survive to harvest at all; every stalk dies. Magic Wheat from America doesn't know beans about Ethiopian drought. The end.

Actually, if the drought arrived in year two and the end came that quickly, in this real-life fairy tale you'd be very lucky, because chances are good you'd still have some of your family-line seed around. It would be much more disastrous if the drought waited until the eighth or ninth year to wipe you out, for then you'd have no wheat left at all, Magic or otherwise. Seed banks, even if they're eleven thousand years old, can't survive for more than a few years on the shelf. If they aren't grown out as crops year after year, they die—or else get ground into flour and baked and eaten—and then this product of a thousand hands and careful selection is just gone, once and for all.

This is no joke. The infamous potato famine or Southern Corn Leaf Blight catastrophe could happen again any day now, in any place where people are once again foolish enough, or poor enough to be coerced (as was the case in Ireland), to plant an entire country in a single genetic strain of a food crop.

While agricultural companies have purchased, stored, and patented certain genetic materials from old crops, they cannot engineer a crop, *ever*, that will have the resilience of land races under a wide variety of conditions of moisture, predation, and temperature. Genetic engineering is the antithesis of variability because it removes the wild card—that beautiful thing called sex—from the equation.

The new magic bullet

This is our new magic bullet: We can move single genes around in a genome to render a specific trait that nature can't put there, such as ultrarapid growth or vitamin A in rice. Literally, we could put a wolf in sheep's clothing. But solving agricultural problems this way turns out to be far less broadly effective than the old-fashioned multigenic solutions derived through programs of selection and breeding. Crop predators evolve in quick and mysterious ways, while gene splicing tries one simple tack after another, approaching its goal the way Wile E. Coyote tries out each new gizmo from Acme only once, whereupon the roadrunner outwits it and Wile E. goes crestfallen back to the drawing board.

Wendell Berry [a conservationist, novelist, and poet], with his reliable wit, wrote that genetic manipulation in general and cloning in particular: ". . . besides being a new method of sheep-stealing, is only a pathetic attempt to make sheep predictable. But this is an affront to reality. As any shepherd would know, the scientist who thinks he has made sheep predictable has only made himself eligible to be outsmarted."

I've heard less knowledgeable people comfort themselves on the issue of genetic engineering by recalling that humans have been pushing genes around for centuries, through selective breeding of livestock and crops. I even read one howler of a quote that began, "Ever since Mendel spliced those first genes. . . ." These people aren't getting it, but I don't blame them—I blame the religious fanatics who kept basic biology out of their grade-school textbooks. Mendel did not *splice* genes, he didn't actually control anything at all; he simply watched peas to learn how their natural system of genetic recombination worked. The farmers who select their best sheep or grains to mother the next year's crop are working with the evolutionary force of selection, pushing it in the direction of their choosing. Anything produced in this way will still work within its natural evolutionary context of variability, predators, disease resistance, and so forth. But tampering with genes outside of the checks and balances you might call the rules of God's laboratory is an entirely different process. It's turning out to have unforeseen consequences, sometimes stunning ones.

The general ignorance of U.S. populations about who controls global agriculture reflects our trust in an assured food supply.

To choose one example among many, genetic engineers have spliced a bacterium into a corn plant. It was arguably a good idea. The bacterium was *Bacillus thuringiensis,* a germ that causes caterpillars' stomachs to explode. It doesn't harm humans, birds, or even ladybugs or bees, so it's one of the most useful pesticides we've ever discovered. Organic farmers have worked for years to expedite the path of the naturally occurring "Bt" spores from the soil, where the bacterium lives, onto their plants. You can buy this germ in a can at the nursery and shake it onto your tomato plants, where it makes caterpillars croak before sliding back into the soil it came from. Farmers have always used nature to their own ends, employing relatively slow methods circumscribed by the context of natural laws. But genetic engineering took a giant step and spliced part of the bacterium's DNA into a corn plant's DNA chain, so that as the corn grew, each of its cells would contain the bacterial function of caterpillar killing. When it produced pollen, each grain would have a secret weapon against the corn worms that like to crawl down the silks to ravage the crop. So far, so good.

But when the so-called Bt corn sheds its pollen and casts it to the wind, as corn has always done (it's pollinated by wind, not by bees), it dusts a fine layer of Bt pollen onto every tree and bush in the neighborhood of every farm that grows it—which is rapidly, for this popular crop, becoming the territory known as the United States. . . . The massive exposure to Bt, now contained in every cell of this corn, is killing off all crop predators except those few that have mutated a resistance to this long-useful pesticide. As a result, those superresistant mutants are taking over, in exactly the same way that overexposure to antibiotics is facilitating the evolution of antibiotic-resistant diseases in humans.

In this context of phenomenal environmental upsets, with even larger ones just offstage awaiting their cue, it's a bit surprising that the objections to genetic engineering we hear most about are the human health effects. It is absolutely true that new combinations of DNA can create proteins we aren't prepared to swallow; notably, gene manipulations in corn unexpectedly created some antigens to which some humans are allergic. The potential human ills caused by ingestion of engineered foods remain an open category—which is scary enough in itself, and I don't mean to minimize it. But there are so many ways for gene manipulation to work from the inside to destroy our habitat and our food systems that the environmental challenges loom as something on the order of a cancer that might well make personal allergies look like a sneeze. If genetically reordered organisms escape into natural populations, they may rapidly change the genetics of an entire species in a way that could seal its doom. One such scenario is the "monster salmon" with genes for hugely rapid growth, which are currently poised for accidental release into open ocean. Another scenario, less cinematic but dangerously omnipresent, is the pollen escaping from crops, creating new weeds that we cannot hope to remove from the earth's face. Engineered genes don't play by the rules that have organized life for three billion years (or, if you prefer, 4,004). And in this case, winning means loser takes all.

Should agribusiness be trusted?

Huge political question marks surround these issues: What will it mean for a handful of agribusinesses to control the world's ever-narrowing seed banks? What about the chemical dependencies they're creating for farmers in developing countries, where government deals with multinational corporations are inducing them to grow these engineered crops? What about the business of patenting and owning genes? Can there be any good in this for the flat-out concern of people trying to feed themselves? Does it seem *safe*, with the world now being what it is, to give up self-sustaining food systems in favor of dependency on the global marketplace? And finally, would *you* trust a guy in a suit who's never given away a nickel in his life, but who now tells you he's made you some *free* Magic Wheat? Most people know by now that corporations can do only what's best for their quarterly bottom line. And anyone who still believes governments ultimately do what's best for their people should be advised that the great crop geneticist Nikolai Vavilov died in a Soviet prison camp.

In light of newer findings, geneticists increasingly concede that gene-tinkering is to some extent shooting in the dark.

These are not questions to take lightly, as we stand here in the epicenter of corporate agribusiness and look around at the world asking, "Why on earth would they hate us?" The general ignorance of U.S. populations about who controls global agriculture reflects our trust in an assured food supply. Elsewhere, in places where people grow more food, watch less TV, and generally encounter a greater risk of hunger than we do, they mostly know what's going on. In India, farmers have persisted in burning to the ground trial crops of transgenic cotton, and they forced their government to ban Monsanto's "terminator technology," which causes plants to kill their own embryos so no viable seeds will survive for a farmer to replant in the next generation (meaning he'd have to buy new ones, of course). Much of the world has already refused to import genetically engineered foods or seeds from the United States. But because of the power and momentum of the World Trade Organization, fewer and fewer countries have the clout to resist the reconstruction of their food supply around the scariest New Deal ever.

Questionable science

Even standing apart from the moral and political questions—if a scientist *can* stand anywhere without stepping on the politics of what's about to be discovered—there are question marks enough in the science of the matter. There are consequences in it that no one knew how to anticipate. When the widely publicized Human Genome Project completed its mapping of human chromosomes, it offered an unsettling, not-so-widely-publicized conclusion: Instead of the 100,000 or more genes that had been expected, based on the number of proteins we must synthesize to be what we are, we have only about 30,000—about the same number as a mustard plant. This evidence undermined the central dogma of how genes work; that is, the assumption of a clear-cut chain of processes leading from a single gene to the appearance of the trait it controls. Instead, the mechanism of gene expression appears vastly more complicated than had been assumed since Watson and Crick discovered the structure of DNA in 1953. The expression of a gene may be altered by its context, such as the presence of other genes on the chromosome near it. Yet, genetic engineering operates on assumptions based on the simpler model. Thus, single transplanted genes often behave in startling ways in an engineered organism, often proving lethal to themselves, or, sometimes, neighboring organisms. In light of newer findings, geneticists increasingly concede that gene-tinkering is to some extent shooting in the dark. Barry Commoner, senior scientist at the Center for the Biology of Natural Systems at Queens College, laments that while the public's concerns are often derided by industry scientists as irrational and uneducated, the biotechnology industry is—ironically—conveniently ignoring the latest results in the field "which show that there are strong reasons to fear the potential consequences of transferring a DNA gene between species."

Recently I heard Joan Dye Gussow, who studies and writes about the energetics, economics, and irrationalities of global food production, discussing some of these problems in a radio interview. She mentioned the alarming fact that pollen from genetically engineered corn is so rapidly contaminating all other corn that we may soon have no naturally bred corn left in the United States. "This is a fist in the eye of God," she said, adding with a sad little laugh, "and I'm not even all that religious." Whatever you believe in—whether God for you is the watchmaker who put together the intricate workings of this world in seven days or seven hundred billion days—you'd be wise to believe the part about the fist.

6

Genetically Modified Organisms Are Contaminating Organic Crops

Ben Lilliston

Ben Lilliston is the communications coordinator for the Institute for Agriculture and Trade Policy, a group that works to assist family farmers in Minneapolis, Minnesota. Lilliston coauthored Genetically Engineered Foods: A Self-Defense Guide for Consumers.

Genetically modified organisms (GMOs), especially corn, contaminate organic crops. GMO corn pollen has drifted onto hundreds of acres of non-GMO farmland. Crops previously grown as organic for years are currently testing positive for traces of GMOs, dramatically decreasing their market value and raising concerns for those wanting organic produce. Organic farmers, who are held responsible for their contaminated crops, seek legislation that would hold seed makers liable instead. Large biotech corporations have faced lawsuits for failing to follow EPA regulations and for inadequately testing GMO crops before releasing them.

Nebraska organic farmer David Vetter has been testing his corn for a new kind of pollution. Situated right in the middle of corn country, Vetter's 280-acre farm is small compared to those of his neighbors. All around him are farmers growing genetically modified corn. And that poses a problem. Corn is an open-pollinating crop. Wind and insects can carry pollen from a few yards to several miles.

[In 2000], Vetter's organic corn tested positive for genetic contamination. "We've been letting customers who buy in bulk know the situation," says Vetter. "Right now, most of it is still sitting in storage on the farm."

Susan Fitzgerald and her husband operate a 1,300-acre farm outside Hancock, Minnesota. [In 2000], Fitzgerald's 100 acres of organic corn showed evidence of genetic contamination, as did her neighbor's organic corn crop. The pollen had traveled more than 120 feet from another neighbor's farm. Instead of selling her organic corn for approximately $4 a bushel, she had to sell her crop on the open market for $1.67.

Vetter and the Fitzgeralds are not alone. Organic farmers are having an increasingly difficult time preventing genetically modified organisms (GMOs) from migrating into their fields. And organic food companies are struggling to ensure the integrity of their products. For consumers who demand organic foods, the alarm bells are ringing.

How much contamination is taking place on organic fields is an unanswered question.

In April [2001], *The Wall Street Journal* tested twenty food products labeled "GMO free" and found that sixteen of them contained at least traces of genetically modified ingredients; five had significant amounts. One of the companies testing positive, albeit with trace amounts, was Nature's Path Foods, the largest organic cereal company in the world.

"We have found traces in corn that has been grown organically for ten to fifteen years," Arran Stephens, president of Nature's Path Foods, told *The New York Times* in June [2001]. "There's no wall high enough to keep that stuff contained."

Biotechnology, utilized primarily on large industrialized farms, splices genes from other plant and animal species into seeds to produce a variety of desired traits, including the ability to withstand exposure to pesticides and even to produce their own pesticides. The most popular genetically modified crops grown in the United States are soybeans, corn, cotton, and canola. Approximately 68 percent of all soybeans and 26 percent of all corn is genetically engineered in the United States, according to June [2001] statistics from the U.S. Department of Agriculture.

But this is counting only those crops that are designed by genetic engineering, not those that are contaminated by it.

The pervasiveness of GMO contamination

How much contamination is taking place on organic fields is an unanswered question.

"For certain crops, it is absolutely pervasive," says David Gould, an organic certification specialist. "Virtually all of the seed corn in this country has at least a trace of GMO contamination and often more. Canola is as bad if not worse. Soy is very problematic, too."

Other crops may also pose risks. Squashes, sugar beets, tomatoes, and potatoes have been approved for bioengineering. "These are not widespread yet," says Gould. "Just give them time, and they'll be a problem, too."

Organic Trade Association Executive Director Katherine D'Matteo says there is some misunderstanding about what organic products are. "We've built the expectation that there is a purity in the world, and even the slightest contamination is a disaster," she says.

"We're seeing traces appearing somewhat more frequently in organic, but we're not seeing an escalation to high percentages," says John Fagan, CEO of Genetic ID, a firm that tests food for many organic and conventional food companies. "If you compare organic with conventional, it is orders of magnitude cleaner."

Genetic contamination can come through the sharing of equipment like combines, elevators, or trucks. And it can also come through seeds. "It is very difficult to find clean seed," says Gould. "Without good seed, we will never be able to produce clean crops."

Jim Riddle, Secretary of the National Organic Standards Board (NOSB), encourages farmers to test all organic seeds to ensure they are free of genetically modified ingredients before planting. Thus far, most organic seeds have not tested positive for this type of contamination. But the American Seed Trade Association recently asked the U.S. Department of Agriculture to establish a tolerance level of 1 percent genetic contamination for seed that is labeled nonmodified.

"It is a pretty good clue that the seed companies can't manage what they are doing when they ask for a tolerance level," says Vetter. "They've come right out and admitted that they can't guarantee non-GMO seed."

Costs and certification concerns

The costs associated with trying to keep organic separated from genetically modified seed are mounting. For farmers, it includes buffer zones, cleaning equipment, inspections of crops and processing facilities, and frequent testing. Seed testing costs on average about $10 a bag. After-harvest testing can cost $400 per sample.

"A real issue at the moment for organic farmers is the increased cost associated with testing," says Bob Scowcroft, with the Organic Farm Research Foundation. "If you're sharing equipment, does the neighbor have to steam clean his combine? What about the truck and elevator if it's multi-use?"

A few years ago, there was little incentive for organic farmers to try to find out whether or not their crops were tainted by genetically modified organisms. Why risk the monetary loss that could result if you discovered your crop was contaminated? But now, most organic farmers are doing some type of testing.

"Your buyers are going to find out," says Vetter. "So farmers are going to have to test."

Organic is the fastest expanding sector of the domestic food business—growing a whopping 20 percent every year since 1990.

Another major concern is the potential for the loss of certification that allows farmers to sell their products as "organic." If an organic crop tests positive, a certifier has to make a judgment call, taking into account the extent of contamination and the farmer's efforts to stop it. The official could pull the farm's certification, or more likely pull organic certification for the contaminated crop, says Riddle, whose board is appointed by the USDA to oversee the implementation of national organic rules.

"I do think the NOSB needs to look at the threshold or rejection level issue," says Riddle. "Organic does not mean chemical free or GMO-free, but it means GMOs are not used in the production. Organic farmers are being penalized by the actions of their neighbors. The tolerance level should be very low."

"The consumers' right to know what is in the food they eat and how it is made must be protected. The FDA is much too busy protecting the profits of the biotech food companies."

While the United States does not have set tolerance levels for organic food, there are some relevant standards in other countries. In Europe, any food with a content of 1 percent or higher of genetically modified ingredients must be labeled as such. In Japan, the rule is 5 percent or higher. In an effort to avoid the labeling requirement, food companies in Japan and Europe are rejecting crops that exceed these thresholds. According to Fagan of Genetic ID, organic companies in Europe and Japan are strict about genetically altered content, with most companies unwilling to accept anything with a 0.1 percent threshold or higher.

Organic food grown in the United States is fast becoming a major export. According to the Organic Trade Association, the United States exports more than $40 million in organic goods to the United Kingdom and an estimated $40 to $60 million to Japan each year. The association estimates that U.S organic exports to Europe are growing by 15 percent per year, and by 30 to 50 percent per year to Japan.

Most other countries expect that organic products coming from the United States will be free of genetically modified ingredients. But that situation could change. In 1999, Europe rejected corn chips manufactured by the Wisconsin company Terra Prima because of genetic contamination. The event cost the company hundreds of thousands of dollars.

Organic is the fastest expanding sector of the domestic food business—growing a whopping 20 percent every year since 1990. There are 7,800 certified organic farms in the United States, up from 6,600 in 1999, according to the Organic Farm Research Foundation. Organic sales will likely increase from an estimated $5.4 billion in 1998 to more than $9 billion in 2001, according to Datamonitor, the food industry analyst.

Organic foods are popular with consumers who prefer natural ingredients and worry about the health hazards of pesticides, synthetic hormones, and other aspects of industrial agriculture.

But if consumers cannot be assured that they are getting organic products free of genetically modified ingredients, the market may diminish.

In 1997, when the U.S. Department of Agriculture proposed national standards that would have considered genetically modified crops to be organic, nearly 300,000 people submitted comments denouncing the plan. A major selling point of organic foods has been that the standard disallows genetically modified organisms. Many organic food companies, like

Nature's Path, Eden Foods, Erewhon, and Gardenburger, tout this claim on their labels.

[In] December [2000] the organic community roundly hailed the conclusion of a tortuous ten-year process to develop national standards for organic food production. While the final standard explicitly excluded genetically modified crops, it was decidedly vague on the issue of contamination. The rules appear to allow some genetic impurities, although they do not specify how much. The rules state, "The presence of a detectable residue of a product of excluded methods [like genetic alterations] alone does not necessarily constitute a violation of this regulation. . . . The unintentional presence of the products should not affect the status of an organic product or operation."

Because of difficulties in ensuring that food is free of genetically modified ingredients, the Food and Drug Administration's new regulations on genetically modified organisms, announced in January [2001], do not allow companies to claim that their products are "GMO-free." Instead, labels will have to say that the food was not produced through bioengineering.

The StarLink incident

The demand for crops that have not been genetically modified has increased dramatically since an unapproved yellow corn called StarLink was found in a taco shell. StarLink, which is genetically engineered to contain the pesticide Bt in every cell, had been approved for animal feed but not for human consumption because of concerns about dangerous allergic reactions. StarLink has since been found in nearly 300 consumer products, and the EPA estimates it will take up to four years before StarLink is completely out of the food system.

"StarLink has made a big difference in terms of understanding how farm contamination can happen," says D'Matteo. "And it showed how the government didn't have controls in place."

In June [2000], StarLink was found in white corn tortilla chips in Florida, according to *The Washington Post.* "The presence of StarLink in a white corn product illustrates how difficult it is to keep genetically modified crops from spreading," reported the *Post* on July 4. "White corn is grown and distributed separately from yellow corn, and industry observers said there are no genetically modified varieties. But they also said it has proven impossible to prevent some commingling of conventional and modified, as well as white and yellow, corn. The mixing, they said, could happen at processing plants, during transportation, and during cross-pollination in fields."

On July 27, the U.S. Environmental Protection Agency announced that it would not establish a tolerance level for StarLink in human foods. This move, opposed by the food industry, effectively bans StarLink from U.S. grocery shelves and mandates testing of corn entering the food supply.

Organic farmers push for legislation

Organic farmers are fighting back.

Many family farm groups throughout the country are interested in pushing for legislation that would clearly identify the seed maker, rather than the farmer, as liable for contamination.

Politicians have introduced bills in the U.S. Congress and more than a dozen states that would require labeling of genetically modified foods and stronger pre-market safety testing requirements. Some of the bills would assign liability to seed companies for damages.

"The consumers' right to know what is in the food they eat and how it is made must be protected. The FDA is much too busy protecting the profits of the biotech food companies," says Representative Dennis Kucinich, Democrat of Ohio.

The Maine legislature passed a bill in May 2000 that would require manufacturers or seed dealers of genetically engineered plants, plant parts, or seeds to provide written instructions to all growers on how to plant, grow, and harvest the crops to minimize potential cross-contamination. The Maine bill is the first of its kind in the country.

But the future integrity of organic products may well be decided in the courtroom. There is no case law related to genetically altered crops, and no laws have passed (although several have been introduced at the state and federal level) assigning liability. In the past, U.S. courts have ruled against pesticide companies for pesticide drift. Farmers hope they would do the same for genetic drift.

Organic farmers also have been active in lawsuits against the Environmental Protection Agency. One suit, filed in October 1999, demands that the EPA withdraw all genetically modified Bt crops, including Star-Link.

Genetically modified Bt crops insert an engineered version of a natural soil bacterium, known as bacillus thurengensis (Bt), into the crop. Organic farmers sometimes used Bt in spray form as a last option to deal with certain pests like the corn borer. But many in the scientific community, and the EPA itself, fear that genetically modified Bt crops will speed up resistance to Bt, thereby rendering the natural spray that organic farmers use ineffective.

A class action lawsuit filed by farmers who did not grow StarLink seeks compensation for lost export markets associated with the scandal. The lawsuit, filed [in] December [2000], alleges that StarLink's manufacturer, Aventis, failed to follow the EPA registration for StarLink corn and neglected to take other precautions to prevent Star Link corn from entering the human food supply chain. As a result, the suit claims, there has been widespread contamination of the U.S. corn crop with StarLink, which has in turn resulted in a loss of export and domestic markets for U.S. corn and a depression in U.S. corn prices. The suit, filed in Illinois, seeks compensatory and punitive damages, as well as injunctive relief requiring Aventis to decontaminate all soil, farming equipment, storage equipment, harvest equipment, transportation facilities, grain elevators, and non-StarLink seed supplies to prevent further contamination.

Another lawsuit, this one against Monsanto, charges, among other things, that the company failed to test genetically modified seeds and crops adequately before releasing them into the food supply. The lawsuit, filed on behalf of farmers by the Washington, D.C., law firm Cohen, Milstein, Hausfeld, & Toll, also charges that Monsanto, together with other companies, formed a global cartel to fix prices on genetically modified seeds and conspired to restrain trade in the GMO corn and soybean market.

Monsanto disputes that the seeds haven't been adequately tested. "This action is another in a series of unsuccessful attempts by veteran antagonists to stop a technology with the potential to improve our environment, increase food production, and improve health," said David Snively, assistant general counsel for Monsanto. "We're confident this suit will be dismissed."

Biotech company sues farmer

The courts won't necessarily work to the advantage of the organic farmers. Many in the organic community are still talking about the Percy Schmeiser case, decided earlier [in 2001] in Canada. Monsanto sued Schmeiser for growing Roundup Ready Canola. Schmeiser claimed that he had not purchased the seed and that pollen had drifted from a neighbor's farm. The Canadian court ruled that it didn't matter whether the material drifted or not. Schmeiser, it said, was infringing on Monsanto's patent rights. The court ordered him to pay $105,000 to Monsanto.

"If U.S. courts allowed biotech companies to sue organic farmers for selling their contaminated crops, organic farmers could be found liable to pay damages to the contaminating companies. In essence, this would amount to requiring organic farmers to pay for the nuisance caused by these biotech companies," wrote San Francisco attorneys Robert Uram and Giselle Vigneron in a recent analysis of the case.

Farmers like Vetter and the Fitzgeralds are going to extraordinary measures to prevent the contamination of their crops. The Fitzgeralds have planted mostly wheat and soybeans, and have tried to use buffers when their neighbors planted genetically modified corn.

Vetter tried to stagger his planting time with his neighbors'. "Because we knew the chances of cross-pollination were great, we tried to offset planting dates with our neighbors," Vetter says. "We hoped that they would plant early, and we waited as long as we could to plant."

Susan Fitzgerald hopes that organic farmers can work together with conventional farmers interested in planting crops that have not been genetically altered.

"We don't want to make enemies," she says. "But we want to defend our right to grow GMO-free crops."

7

Genetically Engineered Foods Will Help Stop World Hunger

Gregory E. Pence

Gregory E. Pence teaches bioethics at the University of Alabama in Birmingham. He was the editor of The Ethics of Food. *He has lectured worldwide and has published articles in the* New York Times, *the* Wall Street Journal, *and* Newsweek.

Genetically engineered crops can solve the world's hunger problem. Traditional farming methods in developing countries, including organic farming, require too much land usage and will not be adequate for the growing world population, which is projected to be 8.3 billion by the year 2025. One crop expected to help in this area is genetically modified "golden rice," a rice that has been enhanced with beta-carotene and iron to improve its nutritional qualities. The development of another biotech rice that increases crop yields up to 35 percent will have a dramatic impact because almost half the world's population depends on rice. Anti-biotech activists should not prevent developing countries such as those in Africa from using this new, potentially life-saving technology.

Will [genetically enhanced crops] allow us to end famine? Norman Borlaug [Nobel Prize–winning plant biologist] answers, "With the technology that we now have available, and with the research information that's in the pipeline and in the process of being finalized to move into production, we have the know-how to produce the food that will be needed to feed the population of 8.3 billion people that will exist in the world in 2025."

Moreover, we may be able to do so without destroying the environment. Borlaug continues, "Modern agriculture saves a lot of land for nature, for wildlife habitat, for flood control, for erosion control, for forest production." He predicts that, "we will be able to produce enough food in 2025 without expanding the area under cultivation very much and

Gregory E. Pence, *Designer Food*. Lanham, MD: Rowman & Littlefield Publishers, Inc., 2002. Copyright © 2002 by Rowman & Littlefield Publishers, Inc. Reproduced by permission.

without having to move into semi-arid or forested mountainous topographies." He concludes, "those . . . values [will remain] that are important to society in general, and especially to the privileged who have a chance to spend a lot of long vacations out looking at nature."

Given the projected increase of the world's population to between 9 and 12 billion over the next fifty years, there seems to be no realistic alternative to Borlaug's ideas. Otherwise, there is not enough land to be used by primitive farming methods to grow the food needed. But to use such land, ideological battles will need to be fought with naturalists who do not want high-yield techniques used in developing countries.

In July 2000, seven academies of science urged rapid acceptance of high-yielding techniques to alleviate world hunger.

In an essay in *Time* magazine in 2000, Microsoft founder Bill Gates argued that only genetically enhanced food could feed the world and also cure its poor people of diseases caused by malnourishment:

> The U.N. estimates that nearly 800 million people around the world are undernourished. The effects are devastating. About 400 million women of childbearing age are iron deficient, which means their babies are exposed to various birth defects. As many as 100 million children suffer from vitamin A deficiency, a leading cause of blindness. Tens of millions of people suffer from other major ailments and nutritional deficiencies caused by lack of food.

> How can biotech help? Biotechnologists have developed genetically modified rice [Golden Rice] that is fortified with beta-carotene—which the body converts into vitamin A—and additional iron, and they are working on other kinds of nutritionally improved crops. Biotech can also improve farming productivity in places where food shortages are caused by crop damage attributable to pests, drought, poor soil and crop viruses, bacteria or fungi.

Shortly after Gates's essay appeared, scientists announced the creation of another new strain of genetically enhanced rice that could boost yields up to 35 percent. This is the kind of Borlaug-type technique that has worked so well before. Already the rice has been field-tested in China, Korea, and Chile. One rice expert said this discovery could be very significant because almost half the world's population depends on rice.

On the other hand, a spokesman for Friends of the Earth retorted that yields could also be increased by the same 35 percent by using "better farming methods." If that means "organic" methods, Borlaug has characterized such a claim as "ridiculous," and even if tried it would require cutting down millions of acres of forests to create the needed land.

Imagine if, a century ago, naturalist organizations had publicly opposed the discovery of vitamins. One can imagine naturalists arguing that if God had meant people to have such things synthesized and taken as pills, He would have made them that way in the Garden of Eden. And many naturalists still oppose use of the pesticide DDT, but let's not forget that DDT wiped out malaria in most of the world.

Africa should decide for itself

Florence Wambugu, a plant geneticist and a native of a small village in Kenya, argues that whether Africa should use GM crops should be up to Africa, not Greenpeace or Europe. She scorns the activities of Greenpeace to prevent introduction of GM crops in Africa: "Greenpeace is a $100 million company. To keep that budget you have to be doing something and doing it well. European people are having opinions forced on them through manipulation and half-truths about how dangerous this [GM] technology is." Moreover, "Some aid workers . . . are being pushed into an anti-GM position from their European office."

Wambugu says that organic farming methods won't stop famine in Africa: "In developed countries, food is getting cheaper because they use more and more technology, but in tropical Africa it is getting more expensive because it is all manually produced. People with a small salary spend almost all of it on food. If we can increase food productivity in rural areas it will bring the price of food down, and generate more money for investment to turn the wider economy around."

She thinks that use of Bt corn [corn injected with a gene of Bacillus thuringiensis] in Africa will create "millions of tons more grain." Her own work has been on the sweet potato, a major food crop in Africa. Monsanto has donated the patent on its genetically enhanced sweet potato to all of Africa via the Kenyan Agricultural Research Institute, headed by Wambugu. The new sweet potato is resistant to a virus that periodically devastates this crop.

GM crops are tools

In July 2000, seven academies of science urged rapid acceptance of high-yielding techniques to alleviate world hunger, including use of genetically enhanced beans, wheat, and rice. They urged people *not to focus on the process of adding a desirable trait to an old crop, but on the actual effects of the new crop to people and environments.*

Right now it's important to grow food in local regions so inchoate food economies can build. Egalitarians are right: such peoples don't need monocultures for export, they need to eat. The above-mentioned academies agreed that biotech so far has only marginally aided small farmers in emerging countries and that such farmers should not have to buy back enhanced versions of their native crops.

But GM crops are tools that can be used for many different purposes. How a tool is used depends on the motives of the person using it: a hammer can build a house or kill someone. So with GM plants. Surely Norman Borlaug and his successors will use this tool to build food houses; surely critics are wrong that this tool itself will not help starving people.

8

Genetically Engineered Crops Will Not Solve the World's Food Problems

GeneWatch

GeneWatch is an independent British organization concerned with the ethics and risks of genetic engineering.

Large multinational chemical companies see the development of genetically engineered (GE) foods as a business opportunity. These corporations are building large international networks to market GE crops throughout the world. Calling themselves "life sciences" industries, they are engineering characteristics into plants that give the company quick paybacks, patenting GE seeds, and producing GE foods mainly for developed countries while ignoring the needs of the developing countries they claim they will save from starvation. In fact, GE crops may act against food security by creating viruses and pests that are immune to the toxins produced by the new crops and by failing to address the underlying causes of hunger, such as war and poverty. Alternative, sustainable systems are already in use focusing on such things as soil, water, and nutrient conservation. Use of these systems in places like India and Honduras has resulted in crops that have tripled in production. Government and industry promotion of GE food as being essential to feed the world's population is a smokescreen for their financial interests.

The proponents of genetically engineered (GE) foods argue that biotechnology is essential to feed the world's growing population and build a sustainable agricultural system.[1] The population, which is currently 5.8 billion, is expected to reach 8 billion by 2020 and 11 billion by 2050.[2,3] The advocates of genetic engineering believe that the increasing demand for food must be met without expanding the amount of land used for agricultural purposes (to protect biodiversity) and by addressing issues of soil erosion, salinisation, overgrazing and pollution of water supplies.[3,4] However, many organisations in less developed countries, aid agencies and environmental groups are less positive about the role genetic engineering can play in solving problems of hunger and tackling environmental degradation.

Business opportunity for multinational companies

The development of GE foods is not being driven by farmers, consumers or less-developed countries but by large multinational chemical companies who have recognised a business opportunity. Six major companies now dominate the production of GE foods worldwide: Monsanto, DuPont, Hoechst, Novartis, Rhöne Poulenc and Zeneca. These now style themselves as the 'Life Sciences' industry with activities which may span food, food additives and pharmaceuticals as well as their more traditional roles of chemical and pesticide production.

Governments of developed countries are also supporting the introduction of GE foods. In 1994, the Biotechnology and Biological Science Research Council (BBSRC) was formed to replace the Agriculture and Food Research Council in Britain, reflecting a change in emphasis in agricultural research. Many representatives of large corporations sit on Research and Strategy Boards of the BBSRC,[5] giving them the ability to influence the research programme. In sharp contrast, consumer and public interest groups (other than the Country Landowners' Association) are given no such opportunity for input.

The European Commission also finances the promotion of GE crops and foods. For example, they have granted £1 million to the so-called 'FACTT [Familiarisation with and Acceptance of Crops incorporating Transgenic Technologies]' project,[6] with a similar amount being contributed by Hoechst and other partners. In effect, the project has become a sales promotion for the GE oilseed rape [canola oil] developed by Hoechst subsidiaries AgrEvo and Plant Genetic Systems to bring about *". . . the creation of familiarity with and acceptance of transgenic crops for farmers, extension organisations, processing industry, regulatory organisations, consumer groups and public interest groups".*[6]

Whilst maximising the patent owner's profits, seed patenting is highly detrimental to small-scale farmers.

The strategy behind GE foods has been primarily to identify characteristics with quick payback such as herbicide resistance. Both Monsanto and AgrEvo will be able to increase sales of their herbicides (glyphosate and glufosinate respectively) by selling their herbicide resistant crops to farmers and tying them in to using their own brands of herbicide. Monsanto's sales of glyphosate have already risen as a result of the introduction of GE Roundup Ready crops in the United States.[7] GE crops with insect resistance and prolonged shelf life have also been included in the first wave of commercialisation.

Disease resistance forms the next major class of GE crops in the research and development (R&D) phase. Most of the virus disease resistance involves using genes from the virus itself to induce resistance in the crops. The mechanisms which underlie this are poorly understood and concerns have been raised that recombinations with other viruses may lead to the production of new strains of disease-causing viruses.[8]

Also in R&D are more herbicide and insect resistant crops and crops with improved characteristics for processing, such as oils with higher concentrations of certain fatty acids and wheat with modified starch content to aid in bread making. Other aims are to improve shelf lives of a wider range of fruits and to make crops resistant to frost and drought. Improving nutritional content is also proposed.

What is noticeable about these developments is that they are mainly being applied to crops of importance to the developed world and fit *"comfortably into modern foods systems that emphasise food processing, consumer niche markets and production efficiency."*[9] There is virtually nothing that is directly relevant to less developed countries and internationally there are just four *"coherent, coordinated"* GE research programmes on Third World crops[2]. Even these are minimally resourced.[10]

The World Bank Panel on Transgenic Crops concluded that technology transfer projects between multinational corporations and less developed countries were so rare that the examples they cited were *"exceptional."*[2] At best, therefore, it seems that applications to such countries will be largely incidental, arising from so-called *"spillover innovations."*[2]

Large companies establish the GE food market

'Life sciences' companies are building large international networks to market their GE crops throughout the world. Monsanto in particular has adopted an extremely aggressive take-over policy in recent years, systematically acquiring seed companies, small cutting edge biotechnology firms and companies holding important genetic resources. Together with strategic agreements to market seeds worldwide, this has established Monsanto in the forefront of GE crop supply and they now dominate the maize, cotton and soybean markets. In their own words, *"The opportunity to expand geographically is creating exponential growth potential for existing products."*[7]

Monsanto and other 'life sciences' companies' control over the world's commercial seed trade means that 40% of this trade is now owned by just ten companies. This market dominance is reinforced through patenting genetic material. Monsanto, for example, has made sweeping patent claims to all GE cotton (US 5,159,135; EP 270355) and brassicas (US 5,188,958; EP 270615; WO 8707299).

Whilst maximising the patent owner's profits, seed patenting is highly detrimental to small-scale farmers, particularly in less developed countries where seed saving and sharing are essential for survival. Now, however, if farmers use patented seed, they will be forced to pay royalties if they keep seed to re-sow in following years. This has been a feature of contracts between Monsanto and farmers in the US who grow their GE herbicide resistant 'Roundup Ready' soybean. They have had to agree not to keep seed for future years and to use only Monsanto's brand version of glyphosate (Roundup) on the crop.

GE crops may act against food security

The 'life sciences' companies claim that their development of GE crops is essential to feed the world and eradicate starvation. However, added to the issues of market control and cost, GE crops may act against food security for other reasons.

Herbicide resistant crops bring risks to the environment, human health and farming[11] and dependency on external inputs, such as branded herbicides, means that they are unlikely to be available to poor, small-scale farmers. Viruses and pests may develop immunity to the toxins produced by disease and insect resistant crops, thus involving farmers in a costly treadmill of replacing varieties with newer, more potent ones.

Research clearly indicates that new weed problems may be created if advantages such as herbicide and disease resistance are passed to wild native flora, allowing them to survive in conditions where they would normally have been unable to do so. Such problems may be even more acute in tropical countries where the majority of the major food crops evolved and related wild species abound.

Furthermore, the analysis expounded by many of those promoting genetic engineering for its ability to feed the world does not address many of the long-term issues nor the seemingly intractable problems and contradictions of today. For example, the problems of agricultural production in Africa have simply been attributed to *"inefficient agricultural systems"* which can be solved by the faster introduction of modern agricultural methods.[3] Such a simplistic dismissal of the problems of food production in Africa such as poverty, war and unpredictable rainfall to sell technological solutions seems cynical in the extreme. It also ignores some of the problems associated with the introduction of modern agriculture.

> On the basis of experiences with earlier efforts to introduce Western agricultural innovations in developing countries, it can be expected that innovations resulting from such a [biotechnology] focus will lead to a decrease of commodity prices and result in a greater instability of the agricultural structure in many developing countries. These developments may lead to a further marginalization of small-scale farmers and to an accelerated migration to the already over crowded cities, where job opportunities are very limited.[12]

This statement comes from a 1990 study which considered the complex nature of agricultural systems in developing countries and the role biotechnology, including genetic engineering, could play. Crucially, it indicates that focusing on a technology rather than the underlying causes of hunger or inability to increase food supplies may exacerbate rather than improve the situation.

A farming system which relied on GE crops could, like other intensive systems, result in short-term yield increases but simultaneously degrade the underlying ecosystems.[13] The yield of Roundup Ready soybeans in the US was apparently 5% higher on average in 1996 and 1997 than that of conventionally bred varieties.[14] However, not all farmers have had positive experiences with GE crops.[15] For instance, the Mississippi Seed Arbitration Council has ruled that Monsanto's GE Roundup Ready cotton failed to perform as advertised [in 1997] and recommended that nearly $2 million be paid to three farmers who had large losses. In Arkansas [the same year], farms growing GE insect resistant cotton had, on average, lower yields than conventional varieties and crops had to be harvested twice rather than once.

Alternatives to genetic engineering

There are alternative systems which could provide food security and increase food production in the future if they were coupled with mechanisms to address inequalities in food supply. Systems which do not involve high levels of input are already widely used, are becoming more sophisticated and can be as productive as high input systems, although they will have differential impacts depending on the part of the world in which they are applied. They also have the added advantage of bringing environmental benefits. These approaches focus on soil, water and nutrient conservation, green manures, raised fields, terracing and integrated pest management. They are most successful in complex and varied agricultural systems and yields can be increased dramatically. In sustainable agriculture projects in Honduras, maize yields have been increased by 300%; in India, yields of millet have been increased by up to 154%; in Burkina Faso [in Africa], sorghum and millet yields increased by 275% (see Note 16 for details of these and many other case studies). These are real, tangible benefits for people now, not PR promises for the future.

There are alternative systems which could provide food security and increase food production in the future if they were coupled with mechanisms to address inequalities in food supply.

Following a comprehensive study of sustainable, low input systems around the world, Jules Pretty of the International Institute of Environment and Development has concluded[16] that introducing such systems would lead to:
 • Stabilised or lower yields in industrialised countries, coupled with substantial environmental improvements;
 • Stabilised or slightly higher yields in Green Revolution lands, with environmental benefits;
 • Substantially increased agricultural yields in complex and diverse lands based mostly on available or local resources.

One of the striking points about these conclusions is that the greatest yield improvements would occur in areas where modern agricultural methods have not yet been introduced, where population increase may be greatest, and where farmers are least likely to be able to afford high input systems. Another important aspect of low input systems is that they tend to be specific to the local conditions, not amenable to globalisation and unprofitable for large corporations. Control is exercised, and benefits felt, at a local level.

However, despite their clear advantages, and in contrast to the promotion of genetic engineering, these alternative approaches to agriculture have been starved of resources and research.

Genetic engineering will perpetuate problems

Although global food production has increased over the past three decades, the benefits have not been evenly reaped. In 1994, food production could have supplied 6.4 billion people (more than the actual population) with an adequate 2,350 calories per day, yet more than 1 billion people do not get enough to eat.[2]

Genetic engineering looks set to perpetuate and intensify many of the problems which have led to present day food insecurity.

Modern agricultural systems have caused serious environmental damage, including pollution and health problems through the use of large amounts of chemical inputs, and the erosion of genetic diversity through reliance on a small number of crops and varieties. Small farmers in less developed countries, unable to afford the expensive inputs of intensive agricultural systems, have been prevented from competing with cheap imports and their livelihoods have been placed at risk. National debt burdens have forced poor countries to focus on cash crops, not staple foods. Genetic engineering looks set to perpetuate and intensify many of the problems which have led to present day food insecurity. Corporate control, products designed for a developed world market, packages of expensive seed and inputs coupled with the potential for further environmental harm as a result of genetic pollution mean any benefits will remain concentrated in developed nations.

The complex issues surrounding food provision are unlikely to be solved by a new technological fix. However, by selling GE foods as a panacea to political and social problems, governments and industry may be able to avoid difficult questions while large multinational corporations can look forward to a prosperous future. The promotion of genetic engineering as an essential prerequisite to feed the world of the future is therefore little more than a smokescreen to drive acceptance of the technology in the developed world and the global aspirations of the companies involved.

Notes

1. See e.g. Monsanto's advertisement in *The Observer*, 2nd August 1998, 'Worrying about starving future generations won't feed them. Food biotechnology will.'
2. Kendall, H.W., Beachy, R., Eisner, T., Gould, F., Herdt, R., Raven, P.H., Schell, J.S. & Swaminathan, M.S. (1997) Bioengineering of crops: report of the World Bank Panel on Transgenic Crops. International Bank for Reconstruction and Development/World Bank: Washington DC.
3. Vasil, I.K. (1998) Biotechnology and food security for the 21st century: a real-world perspective. *Nature Biotechnology* 16: 399–400.

4. Monsanto (1997) Report on Sustainable Development including Environmental, Safety and Health Performance. Monsanto: Missouri.

5. Biotechnology and Biological Sciences Research Council Annual Report 1996–97. BBSRC: Swindon.

6. FACTT: A project to promote Familiarisation with and Acceptance of Crops incorporating Transgenic Technologies in modern agriculture. A demonstration project under Framework Programme IV—European Commission. Paper OCS 8/96, Annex D & Draft Technical Annex—December 19 1995.

7. Monsanto Company 1997 Annual Report.

8. Beringer, J. (1994) Are there unresolved issues regarding the possible generation of new viral pathogens from transgenic plants? Proceedings of the 3rd International Symposium on The Biosafety of Field Tests of Genetically Modified Plants and Microorganisms. November 13–16 1994, Monterey, California.

9. Rissler, J. & Mellon, M. (1996) *The Ecological Risks of Engineered Crops.* MIT Press: Cambridge MA.

10. Puonti-Kaerlas, J. (1998) Cassava Biotechnology. *Biotechnology and Genetic Engineering Reviews* 15: 329–364.

11. GeneWatch (1998) Genetically engineered oilseed rape: agricultural saviour or new form of pollution? GeneWatch: Litton, Buxton, UK.

12. Bunders, J.F.G. (1990) Biotechnology for small-scale farmers in developing countries. Analysis and assessment procedures. VU University Press: Amsterdam.

13. Buttel, F.H. (1997) Some observations on agro-food change and the future of agricultural sustainability movements. In 'Globalising Food. Agrarian questions and global restructuring' D. Goodman & M.J. Watts (eds) Routledge: London.

14. Monsanto (1998) Background. The Roundup Ready soyabean system: sustainability and herbicide use. 11pp.

15. The Gene Exchange. Summer 1998. Union of Concerned Scientists: Washington DC.

16. Pretty, J.N. (1995) Regenerating Agriculture. Earthscan: London.

9

Genetically Engineered Foods Will Benefit Developing Countries

Richard Manning

Richard Manning is an environmental writer who has written six books, one of which is Food's Frontier: The Next Green Revolution.

In developing countries such as India, where subsistence farming is prevalent and new protein-rich crops are desperately needed, genetic food modification is a promising alternative. Instead of allowing profit-oriented biotech corporations to control their agriculture, sophisticated native public-sector scientists are developing new disease-resistant and pest-resistant biotech crops for their countries. Using biotechnology, new food strains can be developed in half the time of traditional plant-breeding techniques, and these crops can be more resistant to potentially devastating insects. This technology is of life and death importance in countries where food shortages are urgent and failed crops have contributed to farmers taking their lives rather than facing the hunger of their families.

Fears that genetically engineered foods will damage the environment have fueled controversy in the developed world. The debate looks very different when framed not by corporations and food activists but by three middle-aged women in saris working in a Spartan lab in Pune, India. The three, each with a doctoral degree and a full career in biological research, are studying the genes of chickpeas, but they begin their conversation by speaking of suicides.

The villain in their discussion is an insidious little worm, a pod borer, which makes its way unseen into the ripening chickpea pods and eats the peas. It comes every year, laying waste to some fields while sparing others. Subsistence farmers expecting a bumper crop instead find the fat pods hollow at harvest. Dozens will then kill themselves rather than face the

Richard Manning, "Eating the Genes," *Technology Review*, vol. 104, July/August 2001, p. 90.

looming hunger of their families. So while the battle wages over "franken-food" in the well-fed countries of the world, here in this Pune lab the arguments quietly disappear.

A generation ago the world faced starvation, and India served as the poster child for the coming plague, occupying roughly the same position in international consciousness then that sub-Saharan Africa does today. The Green Revolution of the 1960s changed all that, with massive increases in grain production, especially in India, a country that now produces enough wheat, rice, sorghum and maize to feed its people. Green Revolution methods, however, concentrated on grains, ignoring such crops as chickpeas and lentils, the primary sources of protein in the country's vegetarian diet. As a consequence, per capita production of carbohydrates from grain in India tripled. At the same time, largely because of population growth, per capita protein production halved.

By allowing researchers to accelerate the development of new, pest-resistant sources of protein, genetic engineering can help fulfill the decades-old promise of the Green Revolution.

The gains in grain yield came largely from breeding plants with shorter stems, which could support heavier and more bountiful seed heads. To realize this opportunity, farmers poured on nitrogen and water: globally, there was a sevenfold increase in fertilizer use between 1950 and 1990. Now, artificial sources of nitrogen, mostly from fertilizer, add more to the planet's nitrogen cycle than natural sources, contributing to global warming, ozone depletion and smog. Add to this the massive loads of pesticides used against insects drawn to this bulging monoculture of grain, and one begins to see the rough outlines of environmental damage the globe cannot sustain.

Scientific communities in the developing world

During this same revolutionary period, India and other countries, including Mexico, Brazil, Chile and Cuba, developed scientific communities capable of addressing many of their own food problems. High on their list is the promise of genetic engineering. In India, researchers have found a natural resistance to pod borers in two other crops, the Asian bean and peanuts, and are trying to transfer the responsible gene to chickpeas. If they are successful, farmers will not only get more protein; they will also avoid insecticides. "The farmer has not to spray anything, has not to dust anything," D.R. Bapat, a retired plant breeder, told me. He need only plant a new seed.

This is the simple fact that makes genetic modification so attractive in the developing world. Seeds are packages of genes and genes are information—exceedingly valuable and powerful information. Biotech corporations can translate that information into profits. Yet when those same packets of power are developed by public-sector scientists in places like India, they become a tool, not for profit, but for quickly distributing important information. There is no more efficient means of spreading information than a seed.

The above argument built only slowly in my mind in the course of researching a book *(Food's Frontier: The Next Green Revolution)* that profiled nine food projects in the developing world, all of which were carried out largely by scientists native to the countries I visited. I expected to encounter low-technology projects appropriate for the primitive conditions of subsistence agriculture in the developing world—and I did. But I also found, in all nine cases, a sophisticated and equally appropriate use of genetic research or genetic engineering.

A lab in Uganda, for example, could not regularly flush its toilets for lack of running water, but could tag DNA. This tagging ability, used in six of the projects I studied, allows researchers to understand and accelerate the breeding of new strains. Typically, an effort to breed a disease- or pest-resistant strain of crop can involve ten years of testing to verify the trait. Using genetic markers cuts that time in half—a difference that gains urgency in countries where test plots are surrounded by poor farmers whose crops are failing for want of that very trait.

In this manner, by allowing researchers to accelerate the development of new, pest-resistant sources of protein, genetic engineering can help fulfill the decades-old promise of the Green Revolution. Our last revolution created a world awash in grain. But if Uganda is to get better sweet potatoes, Peru better mashua and India better chickpeas, then research on those orphan crops will have to catch up rapidly. Biotechnology can help.

Food researchers in developing countries are understandably worried they will be hampered by the controversy over genetically modified foods. Meanwhile, they have a hard time understanding why genetic engineering is the focus of such concern. The gains of the Green Revolution, after all—and for that matter the gains of 10,000 years of agriculture—have in many cases come from mating unrelated species of plants to create something new and better. Every new strain has brought with it the potential dangers now being ascribed with apparent exclusivity to genetic engineering, such as the creation of superresistant pests. Genetic engineering merely refines the tools.

When viewed from labs surrounded by subsistence farmers, where food research is a matter of life and death rather than an intellectual debate, genetic engineering is a qualified good—not without problems and dangers, but still of great promise. Genetic modification of foods becomes a natural extension of the millennia-old practice of plant breeding, less environmentally damaging than many modern alternatives. In the end, DNA is knowledge, which we can hope will build to wisdom, from which we may one day create an agriculture that both supports our population and coexists peacefully with our planet.

10

The FDA Should Require Safety Testing and Labeling of Genetically Engineered Foods

Kathleen Hart

Kathleen Hart, author of Eating in the Dark, *is a journalist who has written about health and the environment for more than fifteen years. She was editor of the* Environmental Health Letter *and covered agriculture and biotechnology for* Food Chemical News.

It is estimated that by mid-1998, the FDA had allowed at least thirty-six gene-altered foods to enter the market without labels and without sufficient scientific study of their potential safety risks. The FDA has granted genetically engineered (GE) foods the status of GRAS, or "generally recognized as safe." However, government scientists have acknowledged that there is no way to assure the safety of GE foods. In May 1998, a coalition of biologists, consumers, rabbis, and Christian clergy sued the U.S. Food and Drug Administration for failing to require safety testing and labeling of GE foods. A key issue in the lawsuit was consumers' right to know about the new genetic material being engineered into their food. Plaintiffs argued that these foods have been measured and detected as being different, yet the FDA maintains that GE foods are virtually identical to conventional ones. GE foods should be rigorously tested using well-designed feeding studies to determine any unintended and unexpected changes that might occur with the consumption of GE foods.

As the groundswell of opposition to genetically modified foods rumbled across Europe in 1998, a small but earnest group of American scientists, religious leaders, and concerned citizens voiced their own discontent with biotech foods. On May 27, 1998, a coalition of rabbis, Christian clergy, biologists, and consumers sued the U.S. Food and Drug Administration for failing to require safety testing and labeling of genetically engineered foods. The lawsuit charged that the FDA's policy endangers public health and violates the religious freedom of individuals who wish to avoid foods that have been engineered with genes from animals and microorganisms.

"Genetic engineering is the most radical technology ever devised by the human brain, yet it has been subjected to far less testing than other new products," Steven Druker, president of the Alliance for Bio-Integrity, the lead plaintiff, said at a press conference in Washington, D.C., to announce the lawsuit. "This lax regulatory process is scientifically unsound and morally wrong."

"Genetic engineering is the most radical technology ever devised by the human brain, yet it has been subjected to far less testing than other new products."

Druker said that Genetic ID, the DNA testing company in Fairfield, Iowa, had tested several soy-based infant formulas in 1997, looking for genetically engineered ingredients. Four baby formulas containing soy ingredients had tested positive.

FDA officials had always maintained that genetically engineered foods were substantially the same as their conventional counterparts. Yet, the plaintiffs argued, if the foods contained new constituents that could be detected and measured, how could government regulators assert they were virtually identical to traditional ones?

"Because of FDA's failure to require labeling, millions of American infants, children and adults are consuming genetically engineered food products each day without their knowledge," charged Andrew Kimbrell, director of the Center for Food Safety in Washington, D.C., another plaintiff in the lawsuit. The center is a small, national membership group set up in the early 1990s to assess the human health and environmental safety of new food technologies. "The FDA has made consumers unknowing guinea pigs for potentially harmful, unregulated food substances," he told the large group of reporters assembled for the morning press conference at the National Press Club, a few blocks from the White House.

Consumers' rights to know

A central issue in the lawsuit involved consumers' rights to know about the new genetic material being engineered into their food. Druker pointed out that because the FDA requires neither labeling nor pre-market approval of gene-altered foods, American consumers have no way of knowing what genetically modified foods are on the market. To the best of anyone's knowledge, he offered, by mid-1998 the FDA had allowed at least thirty-six gene-altered foods to enter interstate commerce without labels and without proper analysis of the potential safety risks associated with their genetic instability.

Most of the thirty-six foods named in the lawsuit were crops that had been genetically engineered either to kill insect pests or to resist a partic-

ular company's brand of herbicides. Foods in this category included genetically modified corn, canola, cotton, potato, and soybean varieties
from AgrEvo, Monsanto, and Novartis. Three other kinds of genetically
engineered foods were listed in the lawsuit as well: virus-resistant papayas
developed by researchers at the University of Hawaii and Cornell University, virus-resistant squash from Seminis Vegetable Seeds and Asgrow Seed
Company, and tomatoes modified for changes in ripening from Monsanto, Agritope, and Zeneca.

Rick Moonen, a New York City chef and owner of two restaurants,
Oceana and Molyvos, explained that his support for the lawsuit grew out
of the obligation he feels to his patrons to know precisely what is in the
foods he serves them. "People come to Oceana because they trust me.
They know that I'm going to source out the highest-quality ingredients in
the market for their dining experience," he said. "By not requiring
mandatory labeling and safety testing of all genetically engineered foods,
the government is taking away my ability to assure customers of the purity of the foods they eat at my restaurant."

A packet of press materials that the litigants had assembled included
a simulated dinner menu of genetically engineered foods. Some of the
transgenic plants and animals on the mock menu were already on the
market. Others were products of federally funded research or crops that
have received permits from the USDA for environmental testing:

APPETIZERS

Fingerling Potatoes with Waxmoth Genes
(served with sour cream from bovine growth hormone
treated cows)
Juice of Tomatoes with Flounder Genes

ENTREE

Braised Pork Loin with Human Growth Genes
Boiled New Potatoes with Chicken Genes
Fried Squash with Watermelon and Zucchini Virus Genes
Toasted Cornbread with Firefly Genes

DESSERT

Rice Pudding with Pea and Bacteria Genes

BON APPETIT!

The menu illustrates plainly enough why the food industry has
adamantly opposed consumer efforts to require labeling on genetically altered products. Given the choice between traditional potatoes and potatoes spiked with waxmoth genes, how many people would select the
potato-insect variety?

Violates free exercise of religion

Religious leaders joining the lawsuit insisted that the FDA's failure to regulate and label gene-altered foods violates the free exercise of religion by

Americans who wish to avoid such foods for religious or moral reasons.

"Everyone who believes the biosphere developed through the purposeful plan of a benevolent God should reject gene-altered food to preserve the dignity of that plan," said Rabbi Harold White, a lecturer in theology at Georgetown University. "Since the dawn of life on Earth, divine intelligence has systematically prevented transfers between widely differing species. Limited human intelligence should not rush to make such unprecedented transfers commonplace."

"Since the dawn of life on Earth, divine intelligence has systematically prevented transfers between widely differing species. Limited human intelligence should not rush to make such unprecedented transfers commonplace."

Steven Druker believes that gene-altered foods violate Halakha, Jewish law—a belief not universally shared by Jewish thinkers. Born and raised in Des Moines, Druker left Iowa in 1964 to attend the University of California at Berkeley. He received a law degree from Berkeley's School of Law in 1972, then practiced law in San Francisco and Los Angeles for a number of years before returning to Iowa. It was in the mid-1990s, while researching a book on ethics and spirituality, that Druker learned about what he calls "the genetic reconfiguration of the food supply."

In 1997 he wrote a paper entitled "Are Genetically Engineered Foods in Accord with Jewish Law?" that examines the finer points of Jewish dietary law. For many observant Jews, the transfer of pig genes into fish or vegetables, for instance, would raise obvious concerns about the resulting genetically altered foods. "Organisms implanted with genes from non-kosher species are themselves non-kosher and must be avoided," Druker concluded. In his interpretation of Jewish law, any rearrangement of genes between species is worrisome. For purposes of food production, he argued that "transposing genes between species that are naturally prevented from cross-breeding is a high-risk venture" that is unsound under Jewish law even when both species are kosher.

The FDA's explanation

FDA officials actually did give some consideration to the ethical and religious implications of inserting animal genes into plants, as I discovered in searching through agency documents. James Maryanski, biotechnology coordinator of the FDA's Center for Food Safety and Applied Nutrition, had a hand in developing the agency's policy on biotech foods in 1992. Maryanski gave the following explanation of why the FDA believes that moving genes from one organism to another should not trouble people for religious, ethical, or moral reasons.

"There are thousands of genes in a plant. When a scientist adds new genes from an animal, it gives that plant several new proteins. But these proteins would not seem to give animal characteristics to the vegetable." One point people are not generally aware of, Maryanski continued, is that "there are genes in humans and animals that are in plants. There is a gene that occurs in rice that also occurs in the human brain. Vegetarians would not avoid rice because of that. Our current view is that these modifications will not result in foods that violate any ethical or religious considerations."

To many, Maryanski's explanation is an unsatisfactory response. The FDA's policy simply declares that genetically engineered foods are substantially equivalent to their traditional counterparts. It does not require that they be labeled unless a gene from a food known to cause allergies has been inserted into the new food plant. But without labels identifying genetically modified foods and ingredients, Americans who wish to avoid them for religious, ethical, or other dietary reasons are denied the right to do so.

No sound scientific basis

The FDA's policy does more than violate religious laws, according to the scientists joining the lawsuit; it also abrogates good scientific principles. Philip Regal, a biologist and professor of ecology, behavior, and evolution at the University of Minnesota at St. Paul, believes "there is no sound scientific basis" for the presumption that genetically engineered foods are equivalent to, and thus as safe as, their natural counterparts. Such a presumption, he contends, "can only be made by systematically ignoring a large body of solid and relevant evidence."

"FDA's claim that we can reliably assure ourselves about the safety of a genetically restructured organism merely by knowing features of the species involved is scientifically unsound."

There is no "general recognition of the safety of genetically engineered foods among those members of the scientific community qualified to make such a judgment," Regal wrote in testimony to the court. He recalled discussions among federal regulators at a policy meeting on biotechnology in Annapolis, Maryland, in 1988, in which "government scientist after scientist acknowledged there was no way to assure the safety of genetically engineered foods. Several expressed the idea that, in order to take this important step of progress, society was going to have to bear an unavoidable measure of risk."

John Fagan, who received his Ph.D. in biochemistry and molecular biology from Cornell University and spent seven years doing research funded by the National Institutes of Health, was another party to the lawsuit. By the mid-1990s, he wrote, in testimony submitted to the court, he had grown concerned that some applications of genetic engineering were outpacing "our knowledge about the intricacies of DNA and its interactions with the rest of the living system" and "our ability to predict and control the outcomes of genetic manipulations." In 1994, taking an ethical stand against what he considered "irresponsible trends in biotechnology," Fagan returned a $613,882 grant to the NIH and withdrew applications worth an additional $1.25 million.

"FDA's claim that we can reliably assure ourselves about the safety of a genetically restructured organism merely by knowing features of the species involved is scientifically unsound," Fagan observed. "The only way to even begin to make a reliable safety assessment of a genetically engineered organism is to rigorously test for unintended and unexpected changes. This at a minimum entails well-designed feeding studies using the whole food, not just an extract of the substance(s) known to be directly produced by the foreign genetic material."

The FDA's authority to oversee the safety of the American food supply is granted by the federal Food, Drug and Cosmetic Act first enacted in 1938. Under that law companies are required to gain pre-market approval from the FDA for food additives. Druker and the other plaintiffs in *Alliance for Bio-Integrity, et al. v. Donna Shalala, et al.,* alleged that the FDA's 1992 policy violates the act because genetically engineered foods involve "material" changes that require labeling.

Generally recognized as safe

The FDA defines a food additive as any substance "the intended use of which results or may reasonably be expected to result, directly or indirectly, in its becoming a component of food . . . and which is not generally recognized as safe" for such use. Food additives include such substances as flavorings, thickeners, artificial sweeteners, and preservatives. Many of the substances added to foods, including spices and food-processing enzymes that have been in the food supply for a long time, are exempted from review by the FDA because they are considered to be "generally recognized as safe," or GRAS.

If a company wants to use a new additive in a food, it must submit a petition to the FDA and provide scientific information showing reasonable certainty that the additive will be safe for the intended food use. These petitions must be publicly disclosed. The FDA has not viewed the genetic materials transferred into bioengineered foods as food additives. Instead, the agency has granted them a blanket status as GRAS.

11

Labeling Genetically Engineered Foods Is Not Feasible

Eli Kintisch

Eli Kintisch is a writer for New Republic *and works on* Forward's *Washington D.C. staff.*

Labeling foods as genetically modified (GM) would cause consumer confusion as well as raise grocery bills. In addition to the fruits and vegetables that are genetically modified, many GM processed foods are already on the market, making it too complicated to label them all. Because GM and non-GM crops are mixed together, some manufacturers are not sure how many, if any, GM organisms (GMOs) are in their food products. The only way to differentiate them would be to have segregated food production lines from field to factory to grocery stores. The cost of such an operation would raise retail prices as much as 10 percent for some foods. A possible solution would be to label foods "GM-free" or "non-GM," and have anti-GM consumers pay for the cost of segregating these foods. Since GM foods have apparently hurt no one so far, the burden of paying extra for labeling foods should rest in the hands of those who think GM foods are a problem.

"Americans have consistently demanded the right to know what's in their food," Senator Barbara Boxer righteously informed participants in a hearing on the labeling of genetically modified (GM) food [in September 2000]. "[W]hy not tell Americans whether the ingredients in their food are natural or genetically engineered?" It's a popular plea. During the presidential campaign, both Al Gore and Ralph Nader promised mandatory labels on GM food. According to a Harris poll, 86 percent of Americans support the idea. "It's the very least that food producers can do," explains Craig Culp of Greenpeace. "People should be able to make informed decisions about what they eat." The argument is simple, commonsensical—and wrong. What consumers have "the right to know" is that mandatory labels for GM food would, in all likelihood, add to shoppers' confusion, as well as to their grocery bills.

When people talk about labeling GM food, they're generally thinking of the vegetable aisle of the supermarket. And if labeling simply meant putting a sticker on the genetically modified tomatoes of the future, it would make sense. But pumped-up fruits and vegetables are just the tip of the GM iceberg. Genetically engineered components—oil from GM soybeans, sugar from GM beets, flour from GM corn—also show up in lots of processed food. In fact, according to Greenpeace, they're present in more than 60 percent of the items on grocery-store shelves. (And, needless to say, they've caused no known health problems at all.) So do you slap labels on these products too? Keep in mind that while the oils, sugars, and flours in question come from genetically modified plants, the ingredients themselves usually are chemically indistinguishable from their non-GM equivalents. Complicating the issue still further are foods—such as beer, yogurt, bread, and cheese—that contain no GM ingredients at all but may be processed by genetically customized enzymes and microorganisms. Should we label these as well?

Given the quiet ubiquity of GM ingredients and processing in the food we already eat, a catchall GM label would be too broad to provide consumers with much guidance.

Given the quiet ubiquity of GM ingredients and processing in the food we already eat, a catchall GM label would be too broad to provide consumers with much guidance. In England, where food manufacturers began voluntarily labeling products containing GM ingredients in 1997, Jackie Dowthwaite of Britain's Food and Drink Federation says that "roughly half of the products in the supermarkets had the labels." And, just to be safe, many manufacturers labeled products as GM even if they weren't sure whether the sugar or flour or oil in them came from GM plants—thus making it impossible for consumers to make informed decisions.

Alternatively, you could label only products that contain GM substances in reasonably large doses—say, 1 percent of the combined ingredients. (This is, in fact, the standard that the 15 European Union countries—including Britain—adopted January [2000].) But this compromise doesn't really address the concerns that prompted the labeling movement in the first place. For activists morally opposed to "meddling" with nature, a 1 percent meddling threshold is still unacceptable. And for those truly concerned about the safety of GM food, it's not much of a safeguard. "We worry about toxic substances in foods that appear at a level of 0.00001 percent," says Joseph Hotchkiss, a professor of food science and toxicology at Cornell University.

The problem of not knowing

What's more, even if labeling were required only for food modified beyond some arbitrary threshold, you'd still have the problem of manufacturers who have no idea how much GM food—if any—their products contain. As in Britain, an American manufacturer of breakfast cereal, for example, may not know where the sugars and oils in its granola come from. Commercial crops such as corn and soybeans are sold in vast quantities, with GM and non-GM plants often inadvertently mixed together. When these crops are processed into oil and other products, things become more confusing still. By the time those products reach the granola manufacturer, there's no telling whether they contain some GM substances or not. After all, soybean and corn oils extracted from GM plants are chemically identical to those from non-GM ones.

The only way to ensure that a given ingredient comes from non-GM sources would be to create separate production lines from field to factory to grocery store. This wouldn't just require additional paperwork and regulatory bureaucracy to keep the GM and non-GM streams segregated. It would require entirely *new* grain bins, trucks, and cleaning procedures to ensure that non-GM crops and products were not "contaminated" by GM varieties as they were harvested, stored, shipped, and processed. Joe Parcell, an economist at the University of Missouri, estimates that segregating GM and non-GM soybeans could add "up to a dollar of segregation costs" to a $5 to $6 bushel of soybeans. A study for the Canadian market by KPMG Consulting projects that such segregation would force a retail-price markup of as much as 10 percent for food containing corn, canola, and soy-based products. For crops such as corn, whose GM varieties can transmit their genes to plants in adjacent fields via pollen, the costs of segregation could be even higher. The Chicago Federal Reserve warned that, in addition to raising prices at the supermarket, the burden of segregating GM and non-GM crops could make "smaller and higher-cost [agriculture] firms less viable." And this burden will only grow as the food industry uses greater and greater varieties of GM ingredients.

Anti-GM activists claim that in the 1990s the food industry used a similar cost "myth" to try to avoid the standardized "Nutrition Facts" labels now on foods. But Cornell's Hotchkiss argues that GM labeling will cost "a lot more." Moreover, the cost will be borne by consumers generally, whether or not they want to avoid GM-labeled food. "The major expense is not putting the label on," says University of Saskatchewan research scientist Alan McHughen, "it's keeping it off."

Label products "GM-free"

All of which points to a solution that would help consumers make informed choices, limit the confusion caused by ubiquitous labels, and raise prices only for those consumers who consider GM food a problem: voluntary, standardized labeling of food without GM ingredients. Only companies that wanted to label their products "GM-free" would have to pay for the attendant segregation, quality control, and verification—and they would pass those costs along to their customers. Toward this end, the Food and Drug Administration, which rejected mandatory labeling [in May 2000], is already drafting guidelines that define when manufacturers can use a "non-GM" label. And the U.S. Department of Agriculture recently announced that food labeled "organic" would not contain GM ingredients, giving consumers an additional way to choose.

A "non-GM" label has the virtue of recognizing that GM-free products are now the exception at your local supermarket, rather than the rule. Consumers willing to pay the extra cost for such specialty products would be able to do so—just as millions of Americans already buy food certified kosher or organic. Those unconcerned about genetic tinkering wouldn't have to fund an extensive labeling regime. And the market would take it from there. If, as anti-GM activists argue, a majority of shoppers would pay extra to avoid GM food, we'd find out. Fueled by consumer demand, the GM-free niche would expand until it began pushing GM food off the shelves. On the other hand, if Americans generally prefer the cost, taste, or whatever of the bioengineered food they've been eating unknowingly (and without ill effect) for years, they'd have that option as well. And evidence suggests that many would take it. Even in the midst of Britain's political firestorm over GM food, tomato paste clearly labeled produced from genetically modified tomatoes sold better than its non-GM competitor in Sainsbury's supermarkets, a British chain. Why? The GM variety, says a company representative, was cheaper.

12

Scientific Arguments Against Biotechnology Are Fallacious

Ronald Bailey

Ronald Bailey is science correspondent for Reason, *a libertarian magazine, and author of* ECOSCAM: The False Prophets of Ecological Apocalypse. *He edited* Earth Report 2000 *and* The True State of the Planet *and has produced several series and documentaries for PBS television and ABC News.*

"Green" (as opposed to "red," or medical) biotechnology is a welcome advancement and promises many diverse benefits. Scientific panels have concluded that genetically engineered (GE) foods are safe to eat, and people have been eating these foods for years without any apparent ill effects. Anti-biotech activists are standing in the way of allowing GE foods to help developing countries, and their arguments against green biotech are insubstantial and fallacious. For example, GE food opponents claim that the Monarch butterfly is threatened by pollen from GE corn crops. Advocates say that tests have shown this fear to be unfounded, and the Monarch butterfly's population has actually increased in the past few years. Further resistance to the new food technology is unethical in light of the starvation it could prevent.

Ten thousand people were killed and 10 to 15 million left homeless when a cyclone slammed into India's eastern coastal state of Orissa in October 1999. In the aftermath, CARE and the Catholic Relief Society distributed a high-nutrition mixture of corn and soy meal provided by the U.S. Agency for International Development to thousands of hungry storm victims. Oddly, this humanitarian act elicited cries of outrage.

"We call on the government of India and the state government of Orissa to immediately withdraw the corn-soya blend from distribution," said Vandana Shiva, director of the New Delhi-based Research Foundation for Science, Technology, and Ecology. "The U.S. has been using the

Orissa victims as guinea pigs for GM [genetically modified] products which have been rejected by consumers in the North, especially Europe." Shiva's organization had sent a sample of the food to a lab in the U.S. for testing to see if it contained any of the genetically improved corn and soy bean varieties grown by tens of thousands of farmers in the United States. Not surprisingly, it did.

"Vandana Shiva would rather have her people in India starve than eat bioengineered food," says C.S. Prakash, a professor of plant molecular genetics at Tuskegee University in Alabama. Per Pinstrup-Andersen, director general of the International Food Policy Research Institute, observes: "To accuse the U.S. of sending genetically modified food to Orissa in order to use the people there as guinea pigs is not only wrong; it is stupid. Worse than rhetoric, it's false. After all, the U.S. doesn't need to use Indians as guinea pigs, since millions of Americans have been eating genetically modified food for years now with no ill effects."

Biotech opponents

Shiva not only opposes the food aid but is also against "golden rice," a crop that could prevent blindness in half a million to 3 million poor children a year and alleviate vitamin A deficiency in some 250 million people in the developing world. By inserting three genes, two from daffodils and one from a bacterium, scientists at the Swiss Federal Institute of Technology created a variety of rice that produces the nutrient beta-carotene, the precursor to vitamin A. Agronomists at the International Rice Research Institute in the Philippines plan to crossbreed the variety, called "golden rice" because of the color produced by the beta-carotene, with well-adapted local varieties and distribute the resulting plants to farmers all over the developing world.

[In June 2000], at a Capitol Hill seminar on biotechnology sponsored by the Congressional Hunger Center, Shiva airily dismissed golden rice by claiming that "just in the state of Bengal 150 greens which are rich in vitamin A are eaten and grown by the women." A visibly angry Martina Mc-Gloughlin, director of the biotechnology program at the University of California at Davis, said "Dr. Shiva's response reminds me of Marie Antoinette, [who] suggested the peasants eat cake if they didn't have access to bread." Alexander Avery of the Hudson Institute's Center for Global Food Issues noted that nutritionists at UNICEF doubted it was physically possible to get enough vitamin A from the greens Shiva was recommending. Furthermore, it seems unlikely that poor women living in shanties in the heart of Calcutta could grow greens to feed their children.

The apparent willingness of biotechnology's opponents to sacrifice people for their cause disturbs scientists who are trying to help the world's poor. At the annual meeting of the American Association for the Advancement of Science [in February 2000], Ismail Serageldin, the director of the Consultative Group on International Agricultural Research, posed a challenge: "I ask opponents of biotechnology, do you want 2 to 3 million children a year to go blind and 1 million to die of vitamin A deficiency, just because you object to the way golden rice was created?"

Vandana Shiva is not alone in her disdain for biotechnology's potential to help the poor. Mae-Wan Ho, a researcher in biology at London's

Open University who advises another activist group, the Third World Network, also opposes golden rice. And according to a *New York Times* report on a biotechnology meeting held in [March 2000] by the Organization for Economic Cooperation and Development, Benedikt Haerlin, head of Greenpeace's European anti-biotech campaign, "dismissed the importance of saving African and Asian lives at the risk of spreading a new science that he considered untested."

Shiva, Ho, and Haerlin are leaders in a growing global war against crop biotechnology, sometimes called "green biotech" (to distinguish it from medical biotechnology, known as "red biotech"). Gangs of anti-biotech vandals with cute monikers such as Cropatistas and Seeds of Resistance have ripped up scores of research plots in Europe and the U.S. The so-called Earth Liberation Front burned down a crop biotech lab at Michigan State University on New Year's Eve in 1999, destroying years of work and causing $400,000 in property damage. Anti-biotech lobbying groups have proliferated faster than bacteria in an agar-filled petri dish. In addition to Shiva's organization, the Third World Network, and Greenpeace, they include the Union of Concerned Scientists, the Institute for Agriculture and Trade Policy, the Institute of Science in Society, the Rural Advancement Foundation International, the Ralph Nader-founded Public Citizen, the Council for Responsible Genetics, the Institute for Food and Development Policy, and that venerable fount of biotech misinformation, Jeremy Rifkin's Foundation on Economic Trends. The left hasn't been this energized since the Vietnam War. But if the anti-biotech movement is successful, its victims will include the downtrodden people on whose behalf it claims to speak.

The apparent willingness of biotechnology's opponents to sacrifice people for their cause disturbs scientists who are trying to help the world's poor.

"We're in a war," said an activist at a protesters' gathering during the November 1999 World Trade Organization meeting in Seattle. "We're going to bury this first wave of biotech." He summed up the basic strategy pretty clearly: "The first battle is labeling. The second battle is banning it."

Later that week, during a standing-room-only "biosafety seminar" in the basement of a Seattle Methodist church, the ubiquitous Mae-Wan Ho declared, "This warfare against nature must end once and for all." Michael Fox, a vegetarian "bioethicist" from the Humane Society of the United States, sneered: "We are very clever little simians, aren't we? Manipulating the bases of life and thinking we're little gods." He added, "The only acceptable application of genetic engineering is to develop a genetically engineered form of birth control for our own species." This creepy declaration garnered rapturous applause from the assembled activists.

Despite its unattractive side, the global campaign against green biotech has had notable successes in recent years. Several leading food companies, including Gerber and Frito-Lay, have been cowed into declaring that they will not use genetically improved crops to make their products. Since 1997, the European Union has all but outlawed the growing and importing of biotech crops and food. [In May 2000] some 60 countries signed the Biosafety Protocol, which mandates special labels for biotech foods and requires strict notification, documentation, and risk assessment procedures for biotech crops. Activists have launched a "Five-Year Freeze" campaign that calls for a worldwide moratorium on planting genetically enhanced crops.

For a while, it looked like the United States might resist the growing hysteria, but in December 1999 the Environmental Protection Agency announced that it was reviewing its approvals of biotech corn crops, implying that it might ban the crops in the future. [In May 2000] the Food and Drug Administration, which until now has evaluated biotech foods solely on their objective characteristics, not on the basis of how they were produced, said it would formulate special rules for reviewing and approving products with genetically modified ingredients. U.S. Rep. Dennis Kucinich (D-Ohio) has introduced a bill that would require warning labels on all biotech foods.

In October [of 2000], news that a genetically modified corn variety called StarLink that was approved only for animal feed had been inadvertently used in two brands of taco shells prompted recalls, front-page headlines, and anxious recriminations. Lost in the furor was the fact that there was little reason to believe the corn was unsafe for human consumption—only an implausible, unsubstantiated fear that it might cause allergic reactions. Even Aventis, the company which produced StarLink, agreed that it was a serious mistake to have accepted the EPA's approval for animal use only. Most proponents favor approving biotech crops only if they are determined to be safe for human consumption.

To decide whether the uproar over green biotech is justified, you need to know a bit about how it works. Biologists and crop breeders can now select a specific useful gene from one species and splice it into an unrelated species. Previously plant breeders were limited to introducing new genes through the time-consuming and inexact art of crossbreeding species that were fairly close relatives. For each cross, thousands of unwanted genes would be introduced into a crop species. Years of "backcrossing"—breeding each new generation of hybrids with the original commercial variety over several generations—were needed to eliminate these unwanted genes so that only the useful genes and characteristics remained. The new methods are far more precise and efficient. The plants they produce are variously described as "transgenic," "genetically modified," or "genetically engineered."

Benefits of biotechnology

Plant breeders using biotechnology have accomplished a great deal in only a few years. For example, they have created a class of highly successful insect-resistant crops by incorporating toxin genes from the soil bacterium Bacillus thuringiensis. Farmers have sprayed B.t. spores on crops as an effective insecticide for decades. Now, thanks to some clever biotechnology, breeders have produced varieties of corn, cotton, and potatoes that make their own insecticide. B.t. is toxic largely to destructive caterpillars such as the European corn borer and the cotton bollworm; it is not harmful to birds, fish, mammals, or people.

Another popular class of biotech crops incorporates an herbicide resistance gene, a technology that has been especially useful in soybeans. Farmers can spray herbicide on their fields to kill weeds without harming the crop plants. The most widely used herbicide is Monsanto's Roundup (glyphosate), which toxicologists regard as an environmentally benign chemical that degrades rapidly, days after being applied. Farmers who use "Roundup Ready" crops don't have to plow for weed control, which means there is far less soil erosion.

One scientific panel after another has concluded that biotech foods are safe to eat, and so has the FDA.

Biotech is the most rapidly adopted new farming technology in history. The first generation of biotech crops was approved by the EPA, the FDA, and the U.S. Department of Agriculture in 1995, and by 1999 transgenic varieties accounted for 33 percent of corn acreage, 50 percent of soybean acreage, and 55 percent of cotton acreage in the U.S. Worldwide, nearly 90 million acres of biotech crops were planted in 1999. With biotech corn, U.S. farmers have saved an estimated $200 million by avoiding extra cultivation and reducing insecticide spraying. U.S. cotton farmers have saved a similar amount and avoided spraying 2 million pounds of insecticides by switching to biotech varieties. Potato farmers, by one estimate, could avoid spraying nearly 3 million pounds of insecticides by adopting B.t. Researchers estimate that B.t. corn has spared 33 million to 300 million bushels from voracious insects.

One scientific panel after another has concluded that biotech foods are safe to eat, and so has the FDA. Since 1995, tens of millions of Americans have been eating biotech crops. Today it is estimated that 60 percent of the foods on U.S. grocery shelves are produced using ingredients from transgenic crops. In April [2000] a National Research Council panel issued a report that emphasized it could not find "any evidence suggesting that foods on the market today are unsafe to eat as a result of genetic modification." Transgenic Plants and World Agriculture, a report issued in July [2000] that was prepared under the auspices of seven scientific academies in the U.S. and other countries, strongly endorsed crop biotechnology, especially for poor farmers in the developing world. "To date," the report concluded, "over 30 million hectares of transgenic crops have been grown and no human health problems associated specifically with the ingestion of transgenic crops or their products have been identified." Both reports concurred that genetic engineering poses no more risks to human health or to the natural environment than does conventional plant breeding.

As U.C.-Davis biologist Martina McGloughlin remarked at June 2000's Congressional Hunger Center seminar, the biotech foods "on our plates have been put through more thorough testing than conventional food ever has been subjected to." According to a report issued in April [2000] by the House Subcommittee on Basic Research, "No product of conventional plant breeding" could meet the data requirements imposed on biotechnology products by U.S. regulatory agencies. "Yet, these foods are widely and properly regarded as safe and beneficial by plant developers, regulators, and consumers." The report concluded that biotech crops are "at least as safe [as] and probably safer" than conventionally bred crops.

In opposition to these scientific conclusions, Mae-Wan Ho points to a study by Arpad Pusztai, a researcher at Scotland's Rowett Research Institute, that was published in the British medical journal *The Lancet* in October 1999. Pusztai found that rats fed one type of genetically modified potatoes (not a variety created for commercial use) developed immune system disorders and organ damage. *The Lancet's* editors, who published the study even though two of six reviewers rejected it, apparently were anxious to avoid the charge that they were muzzling a prominent biotech critic. But *The Lancet* also published a thorough critique, which concluded that Pusztai's experiments "were incomplete, included too few animals per diet group, and lacked controls such as a standard rodent diet. "Therefore the results are difficult to interpret and do not allow the conclusion that the genetic modification of potatoes accounts for adverse effects in animals." The Rowett Institute, which does mainly nutritional research, fired Pusztai on the grounds that he had publicized his results before they had been peer reviewed.

Activists are also fond of noting that the seed company Pioneer Hi-Bred produced a soybean variety that incorporated a gene—for a protein from Brazil nuts—that causes reactions in people who are allergic to nuts. The activists fail to mention that the soybean never got close to commercial release because Pioneer Hi-Bred checked it for allergenicity as part of its regular safety testing and immediately dropped the variety. The other side of the allergy coin is that biotech can remove allergens that naturally occur in foods such as nuts, potatoes, and tomatoes, making these foods safer.

Arguments against labeling

Even if no hazards from genetically improved crops have been demonstrated, don't consumers have a right to know what they're eating? This seductive appeal to consumer rights has been a very effective public relations gambit for anti-biotech activists. If there's nothing wrong with biotech products, they ask, why don't seed companies, farmers, and food manufacturers agree to label them?

The activists are being more than a bit disingenuous here. Their scare tactics, including the use of ominous words such as frankenfoods, have created a climate in which many consumers would interpret labels on biotech products to mean that they were somehow more dangerous or less healthy than old-style foods. Biotech opponents hope labels would drive frightened consumers away from genetically modified foods and thus doom them. Then the activists could sit back and smugly declare that biotech products had failed the market test.

The biotech labeling campaign is a red herring anyway, because the U.S. Department of Agriculture plans to issue some 500 pages of regulations outlining what qualifies as "organic" foods by January, 2001. Among other things, the definition will require that organic foods not be produced using genetically modified crops. Thus consumers who want to avoid biotech products need only look for the "organic" label. Furthermore, there is no reason why conventional growers who believe they can sell more by avoiding genetically enhanced crops should not label their products accordingly, so long as they do not imply any health claims. The FDA has begun to solicit public comments on ways to label foods that are not genetically enhanced without implying that they are superior to biotech foods.

It is interesting to note that several crop varieties popular with organic growers were created through mutations deliberately induced by breeders using radiation or chemicals. This method of modifying plant genomes is obviously a far cruder and more imprecise way of creating new varieties. Radiation and chemical mutagenesis is like using a sledgehammer instead of the scalpel of biotechnology. Incidentally, the FDA doesn't review these crop varieties produced by radiation or chemicals for safety, yet no one has dropped dead from eating them.

The environmentalist case against biotech crops includes a lot of innuendo.

Labeling nonbiotech foods as such will not satisfy the activists whose goal is to force farmers, grain companies, and food manufacturers to segregate biotech crops from conventional crops. Such segregation would require a great deal of duplication in infrastructure, including separate grain silos, rail cars, ships, and production lines at factories and mills. The StarLink corn problem is just a small taste of how costly and troublesome segregating conventional from biotech crops would be. Some analysts estimate that segregation would add 10 percent to 30 percent to the prices of food without any increase in safety. Activists are fervently hoping that mandatory crop segregation will also lead to novel legal nightmares: If a soybean shipment is inadvertently "contaminated" with biotech soybeans, who is liable? If biotech corn pollen falls on an organic cornfield, can the organic farmer sue the biotech farmer? Trial lawyers must be salivating over the possibilities.

The activists' "pro-consumer" arguments can be turned back on them. Why should the majority of consumers pay for expensive crop segregation that they don't want? It seems reasonable that if some consumers want to avoid biotech crops, they should pay a premium, including the costs of segregation.

As the labeling fight continues in the United States, anti-biotech groups have achieved major successes elsewhere. The Biosafety Protocol negotiated in February 2000 in Montreal requires that all shipments of biotech crops, including grains and fresh foods, carry a label saying they "may contain living modified organisms." This international labeling requirement is clearly intended to force the segregation of conventional and biotech crops. The protocol was hailed by Greenpeace's Benedikt Haerlin as "a historic step towards protecting the environment and consumers from the dangers of genetic engineering."

Activists are demanding that the labeling provisions of the Biosafety Protocol be enforced immediately, even though the agreement says they don't apply until two years after the protocol takes effect. Vandana Shiva claims the food aid sent to Orissa after the October 1999 cyclone violated the Biosafety Protocol because it was unlabeled. Greenpeace cited the un-ratified Biosafety Protocol as a justification for stopping imports of American agricultural products into Brazil and Britain. "The recent agreement on the Biosafety Protocol in Montreal" means that governments can now refuse to accept imports of GM crops on the basis of the "precautionary principle," said a February 2000 press release announcing that Greenpeace activists had boarded an American grain carrier delivering soybeans to Britain.

Under the "precautionary principle," regulators do not need to show scientifically that a biotech crop is unsafe before banning it; they need only assert that it has not been proved harmless. Enshrining the precautionary principle into international law is a major victory for biotech opponents. "They want to err on the side of caution not only when the evidence is not conclusive but when no evidence exists that would indicate harm is possible," observes Frances Smith, executive director of Consumer Alert.

Model biosafety legislation proposed by the Third World Network goes even further than the Biosafety Protocol, covering all biotech organisms and requiring authorization "for all activities and for all GMOs [genetically modified organisms] and derived products." Under the model legislation, "the absence of scientific evidence or certainty does not preclude the decision makers from denying approval of the introduction of the GMO or derived products." Worse, under the model regulations "any adverse socioeconomic effects must also be considered." If this provision is adopted, it would give traditional producers a veto over innovative competitors, the moral equivalent of letting candlemakers prevent the introduction of electric lighting.

Concerns about competition are one reason European governments have been so quick to oppose crop biotechnology. "EU countries, with their heavily subsidized farming, view foreign agribusinesses as a competitive threat," Frances Smith writes. "With heavy subsidies and price supports, EU farmers see no need to improve productivity." In fact, biotech-boosted European agricultural productivity would be a fiscal disaster for the E.U., since it would increase already astronomical subsidy payments to European farmers.

The Monarch butterfly question

The global campaign against green biotech received a public relations windfall on May 20, 1999, when *Nature* published a study by Cornell University researcher John Losey that found that Monarch butterfly caterpillars died when force-fed milkweed dusted with pollen from B.t. corn. Since then, at every anti-biotech demonstration, the public has been treated to flocks of activist women dressed fetchingly as Monarch butterflies. But when more-realistic field studies were conducted, researchers found that the alleged danger to Monarch caterpillars had been greatly exaggerated. Corn pollen is heavy and doesn't spread very far, and milkweed grows in many places aside from the margins of cornfields. In the wild, Monarch caterpillars apparently know better than to eat corn pollen on milkweed leaves.

Furthermore, B.t. crops mean that farmers don't have to indiscriminately spray their fields with insecticides, which kill beneficial as well as harmful insects. In fact, studies show that B.t. cornfields harbor higher numbers of beneficial insects such as lacewings and ladybugs than do conventional cornfields. James Cook, a biologist at Washington State University, points out that the population of Monarch butterflies has been increasing in recent years, precisely the time period in which B.t. corn has been widely planted. The fact is that pest-resistant crops are harmful mainly to target species—that is, exactly those insects that insist on eating them.

As one tracks the war against green biotech, it becomes ever clearer that its leaders are not primarily concerned about safety. What they really hate is capitalism and globalization.

Never mind; we will see Monarchs on parade for a long time to come. Meanwhile, a spooked EPA has changed its rules governing the planting of B.t. corn, requiring farmers to plant non-B.t. corn near the borders of their fields so that B.t. pollen doesn't fall on any milkweed growing there. But even the EPA firmly rejects activist claims about the alleged harms caused by B.t. crops. "Prior to registration of the first B.t. plant pesticides in 1995," it said in response to a Greenpeace lawsuit, "EPA evaluated studies of potential effects on a wide variety of non-target organisms that might be exposed to the B.t. toxin, e.g., birds, fish, honeybees, ladybugs, lacewings, and earthworms. EPA concluded that these species were not harmed."

Another danger highlighted by anti-biotech activists is the possibility that transgenic crops will crossbreed with other plants. At the Congressional Hunger Center seminar, Mae-Wan Ho claimed that "GM-constructs are designed to invade genomes and to overcome natural species barriers." And that's not all. "Because of their highly mixed origins," she added, "GM-constructs tend to be unstable as well as invasive, and may be more likely to spread by horizontal gene transfer."

"Nonsense," says Tuskegee University biologist C.S. Prakash. "There is no scientific evidence at all for Ho's claims." Prakash points out that plant breeders specifically choose transgenic varieties that are highly stable since they want the genes that they've gone to the trouble and expense of introducing into a crop to stay there and do their work.

Ho also suggests that "GM genetic material" when eaten is far more likely to be taken up by human cells and bacteria than is "natural genetic material." Again, there is no scientific evidence for this claim. All genes from whatever source are made up of the same four DNA bases, and all undergo digestive degradation when eaten.

Herbicide and pesticide resistance

Biotech opponents also sketch scenarios in which transgenic crops foster superpests: weeds bolstered by transgenes for herbicide resistance or pesticide-proof bugs that proliferate in response to crops with enhanced chemical defenses. As McGloughlin notes, "The risk of gene flow is not specific to biotechnology. It applies equally well to herbicide resistant plants that have been developed through traditional breeding techniques." Even if an herbicide resistance gene did get into a weed species, most researchers agree that it would be unlikely to persist unless the weed were subjected to significant and continuing selection pressure—that is, sprayed regularly with a specific herbicide. And if a weed becomes resistant to one herbicide, it can be killed by another.

As for encouraging the evolution of pesticide-resistant insects, that already occurs with conventional spray pesticides. There is no scientific reason for singling out biotech plants. Cook, the Washington State University biologist, points out that crop scientists could handle growing pesticide resistance the same way they deal with resistance to infectious rusts in grains: Using conventional breeding techniques, they stack genes for resistance to a wide variety of evolving rusts. Similarly, he says, "it will be possible to deploy different B.t. genes or stack genes and thereby stay ahead of the ever-evolving pest populations."

The environmentalist case against biotech crops includes a lot of innuendo. "After GM sugar beet was harvested," Ho claimed at the Congressional Hunger Center seminar, "the GM genetic material persisted in the soil for at least two years and was taken up by soil bacteria." Recall that the Bacillus thuringiensis is a soil bacterium—its habitat is the soil. Organic farmers broadcast B.t. spores freely over their fields, hitting both target and nontarget species. If organic farms were tested, it's likely that B.t. residues would be found there as well; they apparently have not had any ill effects. Even the EPA has conceded, in its response to Greenpeace's lawsuit, that "there are no reports of any detrimental effects on the soil ecosystems from the use of B.t. crops."

The Technology Protection System

Given their concerns about the spread of transgenes, you might think biotech opponents would welcome innovations designed to keep them confined. Yet they became apoplectic when Delta Pine Land Co. and the U.S. Department of Agriculture announced the development of the Technology Protection System, a complex of three genes that makes seeds sterile by interfering with the development of plant embryos. TPS also gives biotech developers a way to protect their intellectual property: Since farmers couldn't save seeds for replanting, they would have to buy new seeds each year.

Because high-yielding hybrid seeds don't breed true, corn growers in the U.S. and Western Europe have been buying seed annually for decades. Thus TPS seeds wouldn't represent a big change in the way many American and European farmers do business. If farmers didn't want the advantages offered in the enhanced crops protected by TPS, they would be free to buy seeds without TPS. Similarly, seed companies could offer crops with transgenic traits that would be expressed only in the presence of chemical activators that farmers could choose to buy if they thought they were worth the extra money. Ultimately, the market would decide whether these innovations were valuable.

If anti-biotech activists really are concerned about gene flow, they should welcome such technologies. The pollen from crop plants incorporating TPS would create sterile seeds in any weed that it happened to crossbreed with, so that genes for traits such as herbicide resistance or drought tolerance couldn't be passed on.

This point escapes some biotech opponents. "The possibility that [TPS] may spread to surrounding food crops or to the natural environment is a serious one," writes Vandana Shiva in her recent book *Stolen Harvest.* "The gradual spread of sterility in seeding plants would result in a global catastrophe that could eventually wipe out higher life forms, including humans, from the planet." This dire scenario is not just implausible but biologically impossible: TPS is a gene technology that causes sterility; that means, by definition, that it can't spread.

If the activists are successful in their war against green biotech, it's the world's poor who will suffer most.

Despite the clear advantages that TPS offers in preventing the gene flow that activists claim to be worried about, the Rural Advancement Foundation International quickly demonized TPS by dubbing it "Terminator Technology." RAFI warned that "if the Terminator Technology is widely utilized, it will give the multinational seed and agrochemical industry an unprecedented and extremely dangerous capacity to control the world's food supply." In 1998 farmers in the southern Indian state of Karnataka, urged on by Shiva and company, ripped up experimental plots of biotech crops owned by Monsanto in the mistaken belief that they were TPS plants. The protests prompted the Indian government to declare that it would not allow TPS crops to enter the country. That same year, 20 African countries declared their opposition to TPS at a U.N. Food and Agriculture Organization meeting. In the face of these protests, Monsanto, which had acquired the technology when it bought Delta Pine Land Co., declared that it would not develop TPS.

Even so, researchers have developed another clever technique to prevent transgenes from getting weeds through crossbreeding. Chloroplasts (the little factories in plant cells that use sunlight to produce energy) have their own small sets of genes. Researchers can introduce the desired genes into chloroplasts instead of into cell nuclei where the majority of a plant's genes reside. The trick is that the pollen in most crop plants don't have chloroplasts, therefore it is impossible for a transgene confined to chloroplasts to be transferred through crossbreeding.

As one tracks the war against green biotech, it becomes ever clearer that its leaders are not primarily concerned about safety. What they really hate is capitalism and globalization. "It is not inevitable that corporations will control our lives and rule the world," writes Shiva in *Stolen Harvest.* In *Genetic Engineering: Dream or Nightmare?* (1999), Ho warns, "Genetic engineering biotechnology is an unprecedented intimate alliance between bad science and big business which will spell the end of humanity as we know it, and the world at large." The first nefarious step, according to Ho, will occur when the "food giants of the North" gain "control of the food supply of the South through exclusive rights to genetically engineered seeds."

Patenting biotechnology

Accordingly, anti-biotech activists oppose genetic patents. Greenpeace is running a "No Patents on Life" campaign that appeals to inchoate notions about the sacredness of life. Knowing that no patents means no investment, biotech opponents declare that corporations should not be able to "own" genes, since they are created by nature.

The exact rules for patenting biotechnology are still being worked out by international negotiators and the U.S. Patent and Trademark Office. But without getting into the arcane details, the fact is that discoverers and inventors don't "own" genes. A patent is a license granted for a limited time to encourage inventors and discoverers to disclose publicly their methods and findings. In exchange for disclosure, they get the right to exploit their discoveries for 20 years, after which anyone may use the knowledge and techniques they have produced. Patents aim to encourage an open system of technical knowledge.

"Biopiracy" is another charge that activists level at biotech seed companies. After prospecting for useful genes in indigenous crop varieties from developing countries, says Shiva, companies want to sell seeds incorporating those genes back to poor farmers. Never mind that the useful genes are stuck in inferior crop varieties, which means that poor farmers have no way of optimizing their benefits. Seed companies liberate the useful genes and put them into high-yielding varieties that can boost poor farmers' productivity.

Amusingly, the same woman who inveighs against "biopiracy" proudly claimed at the Congressional Hunger Center seminar that 160 varieties of kidney beans are grown in India. Shiva is obviously unaware that farmers in India are themselves "biopirates." Kidney beans were domesticated by the Aztecs and Incas in the Americas and brought to the Old World via the Spanish explorers. In response to Shiva, C.S. Prakash pointed out that very few of the crops grown in India today are indigenous. "Wheat, peanuts, and apples and everything else—the chiles that the Indians are so proud of," he noted, "came from outside. I say, thank God for the biopirates." Prakash condemned Shiva's efforts to create "a xenophobic type of mentality within our culture" based on the fear that "everybody is stealing all of our genetic material."

If the activists are successful in their war against green biotech, it's the world's poor who will suffer most. The International Food Policy Research Institute estimates that global food production must increase by 40 percent in the next 20 years to meet the goal of a better and more varied diet for a world population of some 8 billion people. As biologist Richard Flavell concluded in a 1999 report to the IFPRI, "It would be unethical to condemn future generations to hunger by refusing to develop and apply a technology that can build on what our forefathers provided and can help produce adequate food for a world with almost 2 billion more people by 2020."

More advantages of biotechnology

One way biotech crops can help poor farmers grow more food is by controlling parasitic weeds, an enormous problem in tropical countries. Cultivation cannot get rid of them, and farmers must abandon fields infested with them after a few growing seasons. Herbicide-resistant crops, which would make it possible to kill the weeds without damaging the cultivated plants, would be a great boon to such farmers.

By incorporating genes for proteins from viruses and bacteria, crops can be immunized against infectious diseases. The papaya mosaic virus had wiped out papaya farmers in Hawaii, but a new biotech variety of papaya incorporating a protein from the virus is immune to the disease. As a result, Hawaiian papaya orchards are producing again, and the virus-resistant variety is being made available to developing countries. Similarly, scientists at the Donald Danforth Plant Science Center in St. Louis are at work on a cassava variety that is immune to cassava mosaic virus, which killed half of Africa's cassava crop in 1999.

Another recent advance with enormous potential is the development of biotech crops that can thrive in acidic soils, a large proportion of which are located in the tropics. Aluminum toxicity in acidic soils reduces crop productivity by as much as 80 percent. Progress is even being made toward the Holy Grail of plant breeding, transferring the ability to fix nitrogen from legumes to grains. That achievement would greatly reduce the need for fertilizer. Biotech crops with genes for drought and salinity tolerance are also being developed. Further down the road, biologist Martina McGloughlin predicts, "we will be able to enhance other characteristics, such as growing seasons, stress tolerance, yields, geographic distribution, disease resistance, [and] shelf life."

> "Biotechnology isn't going to be a panacea for all the world's ills, but it can go a long way toward addressing the issues of inadequate nutrition and crop losses."

Biotech crops can provide medicine as well as food. Biologists at the Boyce Thompson Institute for Plant Research at Cornell University recently reported success in preliminary tests with biotech potatoes that would immunize people against diseases. One protects against Norwalk virus, which causes diarrhea, and another might protect against the hepatitis B virus which afflicts 2 billion people. Plant-based vaccines would be especially useful for poor countries, which could manufacture and distribute medicines simply by having local farmers grow them.

Shiva and Ho rightly point to the inequities found in developing countries. They make the valid point that there is enough food today to provide an adequate diet for everyone if it were more equally distributed. They advocate land reform and microcredit to help poor farmers, improved infrastructure so farmers can get their crops to market, and an end to agricultural subsidies in rich countries that undercut the prices that poor farmers can demand.

Addressing these issues is important, but they are not arguments against green biotech. McGloughlin agrees that "the real issue is inequity in food distribution. Politics, culture, regional conflicts all contribute to the problem. Biotechnology isn't going to be a panacea for all the world's ills, but it can go a long way toward addressing the issues of inadequate nutrition and crop losses." Kenyan biologist Florence Wambugu argues that crop biotechnology has great potential to increase agricultural productivity in Africa without demanding big changes in local practices: A drought-tolerant seed will benefit farmers whether they live in Kansas or Kenya.

Yet opponents of crop biotechnology can't stand the fact that it will help developed countries first. New technologies, whether reaping machines in the 19th century or computers today, are always adopted by the rich before they become available to the poor. The fastest way to get a new technology to poor people is to speed up the product cycle so the technology can spread quickly. Slowing it down only means the poor will have to wait longer. If biotech crops catch on in the developed countries, the techniques to make them will become available throughout the world, and more researchers and companies will offer crops that appeal to farmers in developing countries.

Activists like Shiva subscribe to the candlemaker fallacy: If people begin to use electric lights, the candlemakers will go out of business, and they and their families will starve. This is a supremely condescending view of poor people. In order not to exacerbate inequality, Shiva and her allies want to stop technological progress. They romanticize the backbreaking lives that hundreds of millions of people are forced to live as they eke out a meager living off the land.

Per Pinstrup-Andersen of the International Food Policy Research Institute asked participants in the Congressional Hunger Center seminar to think about biotechnology from the perspective of people in developing countries:

> We need to talk about the low-income farmer in West Africa who, on half an acre, maybe an acre of land, is trying to feed her five children in the face of recurrent droughts, recurrent insect attacks, recurrent plant diseases. For her, losing a crop may mean losing a child. How can we sit here debating whether she should have access to a drought-tolerant crop variety? None of us at this table or in this room [has] the ethical right to force a particular technology upon anybody, but neither do we have the ethical right to block access to it. The poor farmer in West Africa doesn't have any time for philosophical arguments as to whether it should be organic farming or fertilizers or GM food. She is trying to feed her children. Let's help her by giving her access to all of the options. Let's make the choices available to the people who have to take the consequences.

13

Biotechnology Proponents Suppress Arguments Against Genetically Engineered Foods

Karen Charman

Karen Charman is an investigative journalist specializing in environmental, health, and agricultural reporting.

The biotechnology industry has huge money-making potential, and supporters of the technology are conducting intensive campaigns to promote their business and silence opposition. These campaigns include lobbying government legislators to promote the benefits of biotechnology. As a result, in 2001 the federal budget allocated $310 million for biotech programs compared to $5 million for organic farming. Biotech's dominance in publicly funded universities is growing, and some scientists question the quality of the science driving biotech's advancement. Academic scientists are hesitant to speak out against the new technology for fear of retribution, which some have already experienced. In a survey at Cornell University, nearly half the agricultural, nutrition science, and extension service staff had reservations about the health, safety, and environmental impacts of genetically engineered food crops and doubted that genetically engineered (GE) foods were the answer to global hunger. Nevertheless, corporate funding for university research on GE foods increased fivefold from 1985 to 1995.

When research scientist Arpad Pusztai appeared on British television in August 1998 to talk about his studies of genetically engineered potatoes, he was suspended and later fired from his job at the Rowett Research Institute in Scotland. After a distinguished 36-year career there, his research was terminated, his data seized, and a contract clause was invoked that put his pension in jeopardy. At that point, the contract became a gag order forbidding him to discuss his work or defend himself in

Karen Charman, "Spinning Science into GOLD," *Sierra*, July/August 2001, p. 40. Copyright © 2001 by Karen Charman. Reproduced by permission.

the ensuing six months—during which his scientific reputation was trashed by a fierce cadre of pro-biotech scientists in Britain and around the globe.

What had Pusztai done? With the prior approval of his boss, this world authority on a class of plant compounds called lectins had made the case for food safety testing for all genetically engineered crops. At the time, Pusztai's team was conducting the only independent scientific research in the world designed to test the safety of genetically engineered foods. Originally an enthusiastic supporter of genetic engineering, Pusztai had not expected to find any negative results. Biotech researchers were interested in lectins because of their pesticidal properties and the possibility of inserting genes from the compounds into food crops. So Pusztai was both surprised and alarmed to find that rats fed potatoes genetically engineered with a specific lectin developed disturbing changes in the size and weight of some of their vital organs. He also found evidence of weakened immune systems. A control group of rats fed ordinary potatoes and another fed spuds with the lectin added but not genetically spliced in showed no such results.

Few academics are willing to openly criticize biotechnology for fear of retribution from the biotech boosters.

When the interviewer asked if the lack of safety testing for genetically engineered foods concerned Pusztai, he said it did. When asked if he would eat his own genetically engineered potatoes, Pusztai said he would not, and that he didn't think it was fair to use people as guinea pigs for an untested new technology.

Pusztai's remarks helped galvanize a growing consumer revolt in Europe that has cost the biotech industry dearly. Opposition to genetically engineered foods is now strong there and in many other parts of the world as well. In response, a well-funded and -organized biotech hype machine has emerged to promote biotech food as the solution to world hunger and squelch concerns about its safety. Groups like the U.S.-based Biotechnology Industry Organization (BIO), the industry's main trade and lobbying group, are desperately trying to prevent a similar consumer revolt from happening in the United States. Through sponsorship of scientific research in the nation's universities as well as high-powered lobbying on Capitol Hill, the biotech promoters are doing their best to neutralize critics. Their academic sponsorships channel research away from biotech's potential negative effects, while their closed-door meetings in Washington ensure that consumers don't get adequate food testing or labeling, and organic farmers won't get the regulations they need to keep their crops free of genetic contamination.

Academics discouraged from speaking out

Few academics are willing to openly criticize biotechnology for fear of retribution from the biotech boosters, say biotech skeptics like John Ikerd, a retired agricultural economist from the University of Missouri. In his view, the enormous public resources devoted to biotechnology programs are corporate giveaways that come at the expense of other kinds of research. His own work focused on sustainable agriculture systems for smaller-scale family farms rather than serving the big agribusiness models land-grant universities have been promoting for more than 50 years. Ikerd's type of research is viewed as a threat to corporate agriculture, he says, because it enables farmers to reduce their reliance on the fertilizers, pesticides, and other products that agribusiness companies sell.

Ikerd's candor was not well received at his university. "You become labeled as not a team player, as not one of the trusted members of the faculty," he says. "You are not on committees you used to be on, you're not involved in the leadership of the department, and you don't get write-ups in the university publications. You have to decide before you speak out that you don't care about these repercussions. It's like being a whistleblower."

A survey measuring attitudes toward biotechnology among Cornell University agricultural and nutrition-science faculty and extension staff (who advise farmers) found that nearly half have reservations about the health, safety, and environmental impacts of genetically engineered food crops and doubt they are the answer to global hunger. Strong biotech supporters numbered 37 percent, while 8 percent thought agricultural biotech might have useful applications and help with global hunger but expressed concerns about food safety issues in light of inadequate testing. Though their numbers were fewer, the biotech promoters said they felt very comfortable publicly voicing their views, while the concerned majority did not express that sentiment.

A chilling effect

Ann Clark, a pasture scientist at the University of Guelph in Canada, is among those who have been chastised for expressing reservations. She publicly criticized the lack of food safety testing for transgenic crops. "Within two hours of the press conference releasing the report, my dean had called me unethical," Clark said. "He said I was paid to be a pasture scientist and that I should stick with that. It became quite ugly, because the national media picked it up, and people whose views aren't parallel to mine have used [the dean's remarks] extensively."

Clark has tenure, so she isn't worried about losing her job. But she says her treatment has had a chilling effect on the debate about biotechnology within Canadian universities. "There aren't many academics who will say something if they know their administrators—the people who sit in judgment on their performance—are going to publicly lambaste them," she said. That initial incident has made Clark more determined than ever to raise questions about biotechnology. Besides continuing to speak openly, she has a number of papers on her Web site that discuss the growing dominance of biotech in publicly funded universities and question the quality of the science driving biotech's advancement.

Whether they work directly for biotech companies or receive corporate grants for their work in universities or government research institutes, scientists are generally forbidden to disclose their results because of secrecy clauses in their contracts. Such clauses are likely to proliferate as public support for research and education is replaced by corporate money—a shift that is already well under way. Writing in the March 2000 issue of the *Atlantic Monthly*, Eyal Press and Jennifer Washburn report that corporate funding of university research increased fivefold—from $850 million to $4.25 billion—between 1985 and 1995. By 1997, corporate contributions constituted 40 percent of the overall academic research budget.

Whether they work directly for biotech companies or receive corporate grants for their work in universities or government research institutes, scientists are generally forbidden to disclose their results because of secrecy clauses in their contracts.

Sarah Bantz, a graduate student in agricultural economics at the University of Missouri, is researching private money coming into her university over a 30-year period. To get access to corporate contracts, she had to promise not to reveal any specifics about them. She says that of all the biotech research undertaken at the University of Missouri, only one study is assessing health, safety, or environmental impacts. "Virtually all the research is for product development, one way or another," she says.

Traditionally, universities have been reservoirs of independent thinking where tenured faculty had the academic freedom to analyze and interpret science and its implications for society without pressure from financially interested parties. But as funding ties between private industry and universities grow, the pool of independent research is shrinking. "It would be as if we had to draw our scientists from corporations every time we wanted to convene a body of experts to help us resolve a technical, scientific problem with public-policy implications in society," says Tufts University professor Sheldon Krimsky, an authority on the social implications of science and technology. "Corporations will have much more direction and control over what technologies get introduced and what are considered to be safe and unsafe."

Trouble for organic farmers

Organic farmer David Vetter is facing off with the biotech boosters, too, but they act as if he doesn't exist. Vetter's 280-acre Nebraska farm is a patchwork of sweet corn, popcorn, soybeans, barley, a variety of grasses, legumes, and grazing paddocks for cattle. Visitors, including Fred Kirschenmann, director of the Leopold Center for Sustainable Agriculture at Iowa State University, come away impressed by the care that goes into the operation. "It strikes you when you step out on that farm," says Kirschenmann. "You can see it in the fields. It's just good stewardship."

Vetter may be a good caretaker, but he can't control the wind. Cross-fertilization between corn plants occurs regularly in the Corn Belt as winds carry pollen from field to field. Prior to the first large-scale commercial plantings of genetically engineered crops in 1996, wind pollination did not pose particular problems for organic farmers. Their biggest challenge was trying to keep pesticides from blowing onto their fields. But with the advent of transgenic crops—and growing public disquiet, bolstered by some alarming preliminary data on the health and environmental effects of such crops—farmers like Vetter face a real threat to their livelihood. Vetter has been testing for transgenic contamination since 1998. [In 2000,] he found it.

Transgenic contamination is already widespread: 100 percent of the organic corn samples sent in to be tested from the Midwest [in 2001] showed some degree of genetic contamination, which could result in organic corn growers' losing certification—and probably their markets.

So far, Vetter's customers say they will reluctantly accept a certain amount of transgenic contamination, as long as it stays at very low levels. But Vetter is worried. The loss of the organic market for his corn would hit him hard—its revenue equals the net profit his farm generates. In the meantime, he's saddled with a hefty bill: It cost him $1,500 to test one $4,000 load of corn for contamination. "It's extremely frustrating when you have to pay those kinds of costs, through no fault of your own, because somebody's introduced technology they can't manage," Vetter says.

Years ago, Vetter began planting double rows of pines, with 60 feet of untilled sod in between, creating a buffer zone to protect his crops from pesticides drifting over from neighboring farms. The buffer hasn't prevented transgenic pollution, though, and this time he's adamant that responsibility for his genetically contaminated crop should fall squarely on both those who have introduced bioengineered corn into agriculture and the government agencies that have allowed the widespread use of essentially unregulated genetically engineered crops. "It's now clear that we won't be able to have both genetically engineered and non-GE crops," Vetter says. "As an organic grower, I can no longer guarantee that my crops are GE-free. The only resolution I can see is a ban on biotech crops."

Biotechnology Industry Organization

Michael Phillips, executive director for food and agriculture at the Biotechnology Industry Organization, is trying to make sure that Vetter and farmers like him don't get their way. Phillips and his staff see their task as creating a barrier between biotech critics and Washington legislators, while also working to educate decision-makers on what they claim to be biotech's benefits.

So far, BIO has been extremely successful in its mission. Consumer-oriented biotech legislation—mandatory labeling of genetically engineered ingredients on food packages, which independent consumer polls consistently indicate the public wants, and a pre-market safety approval process for biotech foods—has not gotten far on Capitol Hill. Phillips has said that pre-market approval is "something the industry would never support." He and his colleagues at BIO have also worked to defeat the establishment of any tracking system that could require transgenic seed purchases to be registered. Such registration could establish liability for the kind of contamination that Vetter experienced.

The Biotechnology Industry Organization has nearly 1,000 members, including large and small medical and agricultural biotech companies as well as dozens of universities; several law firms; a number of foreign organizations, including the German Pharmaceutical Association and the Israel

Export Institute; financial brokerage houses like PaineWebber and Lehman Brothers; the global accounting and financial consulting firms Pricewater-houseCoopers and KPMG Peat Marwick; Procter & Gamble; and even government entities such as the Canadian province of Ontario, the Illinois Department of Agriculture, and the New Jersey Economic Development Authority. Membership costs are based on the number of employees as well as revenues. In some cases, annual dues run as high as $230,000.

"Corporations will have much more direction and control over what technologies get introduced and what are considered to be safe and unsafe."

Prior to joining BIO in 1999, Phillips was director of the National Academy of Sciences Board on Agriculture and Natural Resources. When Phillips left the academy for BIO, he was in the middle of directing a study to assess the health and environmental safety of crops genetically engineered to contain pesticides. The revolving door took him swiftly from a group that provides policy-makers with independent scientific advice to one that lobbies on behalf of chemical-intensive agriculture.

Because of the success of such advocacy, Congress has been reluctant to regulate pesticides or promote organic farming and other alternatives to chemical-intensive agriculture. But it does generously fund biotechnology. The 2001 budget allocates $310 million for biotech in agriculture and rural development programs. Federal support for organic farming is less than $5 million.

In agriculture and beyond, biotech has huge money-making potential. Harvard Business School professor Ray A. Goldberg predicts the new genetic technologies will revolutionize the global economy by turning traditionally distinct industry sectors—agriculture, health care, energy, and computing—into one gargantuan lifescience industry with "virtually unlimited commercial [patent and ownership] possibilities." Asked to quantify the value of future biotech markets, Goldberg says he had been thinking it could reach $16 trillion. But then he changed his mind, saying that there really isn't any way to put a number on future markets for "virtually everything."

Biotech on Capitol Hill

In autumn 1999, Phillips's organization held "Biotechnology School," weekly or bi-weekly meetings between BIO staff and members of the House Committee on Agriculture and their staffs. At these sessions, BIO taught its congressional pupils what biotechnology is, how it's being used in food and agriculture, and where the science is leading. According to one congressional source who requested anonymity, BIO's school exemplified "typical industry access" to Congress that citizen groups simply don't have. "The agriculture committee is going to control the biotech debate in Congress, and they basically said, 'Come on in BIO, here's everybody you need to lobby. And you can do it every week or as much as you want,'" the source said. "This offer is not extended to environmental or food-safety groups—no way, no how."

BIO has also set up congressional biotechnology caucuses—one in the House and one in the Senate—that work with the industry to advance its issues. Adam Kovacevich, a spokesperson for Cal Dooley (D-Calif.), one of the four co-chairs of the House Biotech Caucus, describes the 65-member group as a "forum for advocacy" that "educates fellow members of Congress on the positive implications of biotechnology." Two of the co-chairs, one Republican and one Democrat, sit on the House Agriculture Committee, and the two others, also one from each party, are on the House Commerce Committee, which has jurisdiction over medical applications of biotechnology.

Though the caucus is not promoting any particular bill, it alerts caucus members to any legislative or regulatory activity that could affect biotechnology. This activity clearly helps keep legislators in the biotech camp. In the 2001 session of Congress, a bill requiring labels on genetically engineered foods was introduced by Representative Dennis Kucinich (D-Ohio). Only one member of the biotech caucus, Mark Udall (D-Colo.), supported the ill-fated bill. Udall's district includes the environmentally aware community of Boulder as well as an area with a lot of biotech companies, says Jennifer Barrett, a legislative assistant in his office. "He cosponsored the labeling bill because he's concerned that consumers should have all the information they need about the food they are eating," she says.

The caucus also organizes forums where invited experts brief members on various biotech issues. Richard Caplan, who works on biotech issues for the U.S. Public Interest Research Group, contacted Dooley's office, offering to present his perspective on biotech food issues. His offer was ignored.

An aid to one of the leaders of the biotech caucus defended the group's orientation. "We're primarily interested in getting out the facts and the science," he said. "We're trying to make this a debate that's based not so much on passion and assumptions but on the actual science." But without the voices of researchers like Arpad Pusztai, farmers like David Vetter, and public-interest advocates like Richard Caplan, one wonders whether it's a debate at all or just nonstop communiques from the biotech hype machine.

Organizations to Contact

The editors have compiled the following list of organizations concerned with the issues debated in this book. The descriptions are derived from materials provided by the organizations. All have publications or information available for interested readers. The list was compiled on the date of publication of the present volume; names, addresses, phone and fax numbers, and e-mail addresses may change. Be aware that many organizations take several weeks or longer to respond to inquiries, so allow as much time as possible.

The Alliance for Better Foods
700 13th St. NW, Suite 800, Washington, DC 20005
(202) 783-4573 • fax: (202) 783-4574
website: www.betterfoods.org

The Alliance for Better Foods supports biotechnology as a safe way to provide a more abundant, nutritious, and higher quality food supply. It supports fact-based discussion about development in plant biotechnology.

Alliance for Bio-Integrity
406 W. Depot Ave., Fairfield, IA 52556
(641) 472-5554
e-mail: info@biointegrity.org • website: www.bio-integrity.org

The Alliance for Bio-Integrity is a nonprofit, nonpolitical organization dedicated to the advancement of human and environmental health through sustainable and safe technologies. It aims to inform the public about technologies and practices that negatively impact health and the environment and to inspire broad-based, responsible action that helps correct the problems and uphold the integrity of the natural order. The alliance's initial objective is a more rational and prudent policy on genetically engineered foods.

BioDemocracy Campaign
6114 Hwy. 61, Little Marais, MN 55614
(218) 226-4164 • fax: (218) 226-4157
website: www.purefood.org

The BioDemocracy Campaign promotes food safety, organic farming, and sustainable agriculture practices. It provides information on the hazards of genetically engineered food, irradiated food, and food grown with toxic sludge fertilizer. It organizes boycotts and protests around these issues and publishes *BioDemocracy News*. Its website includes many fact sheets and articles on genetically modified foods.

Biotechnology Industry Organization (BIO)
1225 Eye St. NW, Suite 400, Washington, DC 20005
(202) 962-9200
e-mail: mlyons@bio.org • website: www.bio.org

BIO represents biotech companies, academic institutions, state biotech centers, and related organizations that suppport the use of biotechnology in improving health care, agriculture, and efforts to clean up the environment. It works to educate the public about biotechnology and responds to concerns about the safety of genetic engineering.

Campaign to Label Genetically Engineered Foods
PO Box 55699, Seattle, WA 98155
(425) 771-4049
e-mail: label@thecampaign.org • website: www.thecampaign.org

The mission of the Campaign to Label Genetically Engineered Foods is to create a national grassroots consumer campaign for the purpose of lobbying Congress and the president to pass legislation that will require the labeling of genetically engineered foods in the United States.

Center for Ethics and Toxics (CETOS)
39120 Ocean Dr., Suite C-2-1, PO Box 673, Gualala, CA 95445
(707) 884-1700
e-mail: cetos@cetos.org • website: www.cetos.org

CETOS is a nonprofit environmental group working on the issues of toxicity from nonconsensual chemical exposure, radical transformation in agriculture with the use of biotechnology, and restructuring humankind through the use of genetic technology at early stages of life. It provides educational information through articles, books, and speaking engagements and acts directly to encourage new policies designed to protect populations at risk.

Center for Food Safety (CFS)
660 Pennsylvania Ave. SE, Suite 302, Washington, DC 20003
(202) 547-9359
e-mail: office@centerforfoodsafety.org • website: www.centerforfoodsafety.org

CFS is a public interest and environmental advocacy organization that works to address the impacts of the food production system on human health, animal welfare, and the environment. It works to achieve its goals through grassroots campaigns, public education, media outreach, and litigation. Among its goals are ensuring the testing, labeling, and regulation of genetically engineered foods and preserving strict national organic food standards.

Center for Global Food Issues (CGFI)
PO Box 202, Churchville, VA 24421-0202
(540) 337-6354 • fax: (540) 337-8593
e-mail: cgfi@rica.net • website: www.cgfi.org

CGFI is a proponent of genetically engineered foods that analyzes and conducts research on agriculture and the environmental concerns surrounding food and fiber production. Its goals are to promote free trade in agricultural products for both economic efficiency and environmental conservation, to combat efforts to limit technological innovation in agriculture, and to heighten awareness of the connection between agricultural productivity and environnental conservation.

Council for Agricultural Science and Technology (CAST)
4420 W. Lincoln Way, Ames, IA 50014-3447
(515) 292-2125 • fax: (515) 292-4512
e-mail: cast@cast-science.org • website: www.cast-science.org

CAST is an industry-based organization that dispenses "science-based" information on food and agricultural policy regionally, nationally, and internationally.

Council for Biotechnology Information
PO Box 34380, Washington, DC 20043-0380
(202) 467-6565
website: www.whybiotech.com

The Council for Biotechnology Information consists of the seven leading biotech companies and two trade associations (Biotechnology Industry Organization and CropLife). It is attempting to create a new communications initiative to improve understanding and acceptance of biotechnology.

Council for Responsible Genetics (CRG)
5 Upland Rd., Suite 3, Cambridge, MA 02140
(617) 868-0870 • fax: (617) 491-5344
e-mail: crg@gene-watch.org • website: www.gene-watch.org

CRG is a national nonprofit organization of scientists, environmentalists, public health advocates, physicians, lawyers, and other concerned citizens. It encourages informed public debate about the social, ethical, and environmental implications of new genetic technologies and advocates the socially responsible use of these advances.

Environmental Defense Fund (EDF)
257 Park Ave. S., New York, NY 10010
(800) 684-3322
e-mail: members@environmentaldefense.org • website: www.edf.org

EDF is a leading national nonprofit organization representing more than three hundred thousand members. It attempts to link science, economics, and law to cost-effective solutions to environmental problems. EDF is dedicated to protecting the environmental rights of all people, including access to clean air and water, healthy and nourishing food, and a flourishing ecosystem.

Food and Drug Administration (FDA) of the United States
5600 Fishers Ln., Rockville, MD 20857-0001
website: www.fda.gov

The FDA is the agency of the federal government responsible for ensuring the safety of America's food and drugs. It approves products for the market and monitors their safety after they are in use.

Food Industry Environmental Network (FIEN)
33 Falling Creek Ct., Silver Spring, MD 20904
(301) 384-8287 • fax: (301) 384-8340
e-mail: JLC@fien.com • website: www.fien.com

FIEN is an information service on genetically modified foods, biotechnology, and the environment. It covers safety, pesticides, hazardous substances, agricultural research, and other environmentally related issues.

Friends of the Earth (FoE)
1025 Vermont Ave. NW, Washington, DC 20005-6303
(202) 783-7400 • fax: (202) 783-0444
e-mail: foe@foe.org • website: www.foe.org

FoE is a radical, national environmental organization dedicated to the protection of the diversity of the planet for future generations. It is based in England, with affiliates in sixty-three countries.

National Center for Food and Agricultural Policy (NCFAP)
1616 P St. NW, Washington, DC 20036
(202) 328-5048 • fax: (202) 328-5133
e-mail: ncfap@ncfap.org • website: www.ncfap.org

NCFAP is a private, nonprofit, nonadvocacy research organization that conducts studies in four program areas: biotechnology, pesticides, international trade and development, and farm and food policy.

Sustain: The Environmental Information Group
920 N. Franklin St., Suite 301, Chicago, IL 60610-3121
(312) 951-8999 • fax: (312) 951-5696
website: www.sustainusa.org

Sustain is a nonprofit organization that promotes a healthy, sustainable environment through innovative communications strategies. It aims to mobilize the public against biotechnology, especially by organizing rallies outside hearings at public agencies.

Syngenta AG
PO Box CH 4002, Basel, Switzerland
website: www.syngenta.com

Syngenta is a world-leading agribusiness that ranks first in crop protection and third in the high-value commercial seeds market. Sales in 2001 were approximately $6.3 billion (U.S.). Syngenta employs more than twenty thousand people in over fifty countries. The company is committed to sustainable agriculture through innovative research and technology.

Union of Concerned Scientists (UCS)
2 Brattle Sq., Cambridge, MA 02238-9105
(617) 547-5552 • fax: (617) 864-9405
website: www.ucsusa.org

UCS is an independent, nonprofit alliance of fifty thousand concerned citizens and scientists across the country. Through the Sound Science Initiative, two thousand scientists provide facts on environmental science to government and the media.

Websites

Center for Food and Agricultural Research (CFFAR)
www.cffar.org

CFFAR is a think tank that supports genetically modified foods. It is a public policy and research coalition dedicated to exploring and understanding health, safety, and sustainability issues associated with food and fiber production.

Citizens for Health
www.citizens.org

This group is a nonprofit, grassroots, consumer advocacy group that champions public policies that empower individuals to make informed health choices.

CropChoice
www.cropchoice.com

CropChoice is a coalition of food groups opposed to genetically modified foods. It is an information source for American farmers about genetically modified crops, alternative management options, and profitability.

Ecological Farming Association (EFA)
www.eco-farm.org

The Ecological Farming Association is a nonprofit educational organization and an alternative farming systems information center that promotes ecologically sound agriculture.

GeneLetter
www.geneletter.org

GeneLetter is the leading online magazine of genetics, society, and culture. It offers daily news and monthly features that explore scientific, medical, and bioethical issues surrounding genetics.

Life Sciences Knowledge Center
www.biotechknowledge.com

This site is sponsored by the leading biotech corporation, Monsanto, and provides news items, technical reports, and other documents connected with biotechnology.

Northern Light
http://special.northernlight.com

This website is compiled by Cambridge-based researchers and librarians and is a comprehensive source for information and links on genetically modified food. It provides links to information available electronically from several different perspectives.

Pew Initiative on Food and Biotechnology
http://pewagbiotech.org

The Pew Initiative on Food and Biotechnology is an independent and objective source of information on genetically modified food and agricultural biotechnology for the public, media, and policy makers. The initiative is neither for nor against agricultural biotechnology but is committed to providing information and encouraging debate and dialogue.

Physicians and Scientists for Responsible Application of Science and Technology (PSRAST)
www.psrast.org

PSRAST is a global scientific organization with no political, ideological, or commercial ties. Its purpose is to assess the safety of new technologies in an impartial way. It is not against genetic engineering, but is strongly against any application before health and environmental safety have been established.

Bibliography

Books

Gordon Conway	*The Doubly Green Revolution: Food for the 21st Century.* Ithaca, NY: Cornell University Press, 1997.
Kristin Dawkins	*Gene Wars: The Politics of Biotechnology.* New York: Seven Stories Press, 2002.
Michael W. Fox	*Beyond Evolution: The Genetically Altered Future of Plants, Animals, the Earth, and Humans.* New York: Lyon Press, 1999.
Michael W. Fox	*Eating with Conscience: The Bioethics of Food.* Troutdale, OR: NewSage Press, 1997.
Mae-Wan Ho	*Genetic Engineering: Dream or Nightmare?* Bath, England: Gateway Books, 1998.
Craig Holdredge	*Genetics and the Manipulation of Life: The Forgotten Factor of Context.* Hudson, NY: Lindisfarne Press, 1996.
Alan Holland and Andrew Johnston, eds.	*Animal Biotechnology and Ethics.* London: Chapman and Hall, 1998.
Jack Kloppenburg et al.	*Does Technology Know Where It's Going?* Edmonds, WA: Edmonds Institute, 1996.
Brewster Kneen	*Farmageddon: Food and the Culture of Biotechnology.* Gabriola Island, Canada: New Society of Publishers, 1999.
Sheldon Krimsky and Roger Wrubel	*Agricultural Biotechnology and the Environment.* Chicago: University of Illinois Press, 1996.
Bill Lambrecht	*Dinner at the New Gene Café.* New York: St. Martin's Press, 2001.
Mark Lappe and Britt Baily	*Against the Grain: Biotechnology and Corporate Takeover of Your Food.* Monroe, ME: Common Courage Press, 1998.
Alan McHughen	*Pandora's Picnic Basket: The Potential and Hazards of Genetically Modified Foods.* Oxford, England: Oxford University Press, 2000.
Margaret Mellon and Jane Rissler	*The Ecological Risks of Engineered Crops.* Cambridge, MA: MIT Press, 1996.
Margaret Mellon and Jane Rissler	*Now or Never: Serious New Plans to Save a Natural Pest Control.* Cambridge, MA: Union of Concerned Scientists, 1998.
Stephen Nottingham	*Eat Your Genes: How Genetically Modified Food Is Entering Our Diet.* New York: Zed Books, 1998.

Jeremy Rifkin	*The Biotech Century: Harnessing the Genes and Remaking the World*. New York: Tarcher/Putnam, 1998.
Vandana Shiva	*Biopiracy: The Plunder of Nature and Knowledge*. Boston: South End Press, 1997.
Paul B. Thompson	*Food Biotechnology in Ethical Perspective*. London: Chapman and Hall, 1997.
Laura and Robin Ticciati	*Genetically Engineered Foods: Are They Safe? You Decide*. New York: Keats, 1998.
Jon Turney	*Frankenstein's Footsteps: Science, Genetics, and Popular Culture*. New Haven, CT: Yale University Press, 1998.
Johannes Wirz and Lammerts Van Bueren, eds.	*The Future of DNA*. Dordrecht, The Netherlands: Kluwer Academic Publishers, 1997.

Periodicals

Laurent Belsie	"No Bumper Crop of Genetically Altered Plants," *Christian Science Monitor*, August 31, 2001.
Burton Bollag	"Public Pressure Puts a Damper on Biotechnology Research in Europe," *Chronicle of Higher Education*, April 14, 2000.
Mary Ellen Butler	"Mandatory Labeling of Biotech Foods Still up for Debate," *Food Chemical News*, October 2, 2000.
Bryan Christie	"Scientists Call for a Moratorium on Genetically Modified Foods," *British Medical Journal*, February 20, 1999.
Gordon Conway	"The Voice of Reason in the Global Food Fight," *Fortune*, February 21, 2000.
CQ Weekly	"The Origin of Engineered Food," April 28, 2001.
Economist	"Butterfly Balls," September 22, 2001.
Economist	"Feeding the Five Billion," November 10, 2001.
Pamela Emanoli	"So What Do You Think?" *Human Ecology*, Fall 2000.
Martin Enserink	"Ag Biotech Moves to Mollify Its Critics," *Science*, November 2001.
Doug Farquhar and Crystal Biggerstaff	"Playing God with Potatoes," *State Legislatures*, January 2002.
Nancy Fitzgerald and Nicole Dyer	"Superfood or Double Trouble? Genetically Modified Foods Cook Up a Sizzling Debate," *Scholastic Choices*, February 2002.
Peter Ford	"Pate, Bonhomie, and a Slap at Engineered Food," *Christian Science Monitor*, August 31, 2001.
Julie Forster and Geri Smith	"A Genetically Modified Comeback," *Business Week*, December 24, 2001.
Timothy Gower	"Should You Fear the New Foods?" *Better Homes and Gardens*, November 2000.

Peter Gwin   "Genetically Modified Crops," *Europe*, June 2001.

Kathleen Hart   "Consumers More Likely to Hear Negative GM News,"
                            Food Chemical News, July 2, 2001.

David Hosansky   "Biotech Foods: Trust Is on the Table," *CQ Weekly*, April
                            28, 2001.

Leighton Jones   "Genetically Modified Foods," *British Medical Journal*,
                            February 27, 1999.

Andrea Knox   "Groups Urge Testing Regulation of Genetically Modi-
                            fied Foods," Knight-Ridder/Tribune News Service, July
                            19, 2000.

Richard Manning   "Eating the Genes," *Technology Review*, July/August 2001.

Nutrition Action   "Genetically Engineered Foods," November 2001.
Healthletter

David Orgel   "It's Time to Put GM Foods on Consumers' Radar
                            Screens," *Supermarket News*, March 4, 2002.

Robert Paarlberg   "The Global Food Fight: Genetically Modified Foods at
                            Home and Abroad," *Montana Business Quarterly*, Autumn
                            2001.

Bob Phelps   "Genetic Engineering: Freeze It for Five Years," *Habitat
                            Australia*, December 2000.

Paul Raeburn   "Warning: Biotech Is Hurting Itself," *Business Week*,
                            December 20, 1999.

Paul Rauber   "Eater Beware!" *Sierra*, July/August 2000.

Ziauddin Sarder   "Thank You for the Genes We Eat," *New Statesman*, Janu-
                            ary 15, 2001.

Michael Specter   "The Pharmageddon Riddle: Did Monsanto Just Want
                            More Profits, or Did It Want to Save the World?" *New
                            Yorker*, April 10, 2000.

David Stipp   "The Voice of Reason in the Global Food Fight," *Fortune*,
                            February 21, 2000.

Cass R. Sunstein   "Is Nature Good?" *New Republic*, October 23, 2000.

Colin Tudge   "Why We Don't Need GM Foods," *New Statesman*, Febru-
                            ary 19, 1999.

Index